Programming the Semantic Web

Programming the Semantic Web

Toby Segaran, Colin Evans, and Jamie Taylor

O'REILLY®

Beijing · Cambridge · Farnham · Köln · Sebastopol · Tokyo

Programming the Semantic Web

by Toby Segaran, Colin Evans, and Jamie Taylor

Published by O'Reilly Media, Inc., 1005 Gravenstein Highway North, Sebastopol, CA 95472.

O'Reilly books may be purchased for educational, business, or sales promotional use. Online editions are also available for most titles (*http://my.safaribooksonline.com*). For more information, contact our corporate/institutional sales department: 800-998-9938 or *corporate@oreilly.com*.

Editor: Mary E. Treseler
Production Editor: Sarah Schneider
Copyeditor: Emily Quill
Proofreader: Sarah Schneider

Indexer: Seth Maislin
Cover Designer: Karen Montgomery
Interior Designer: David Futato
Illustrator: Robert Romano

Printing History:
  July 2009:   First Edition.

ISBN: 978-0-596-15381-6

[LSI]   [2010-12-17]

1292003004

Table of Contents

Part II. Standards and Sources

Part III. Putting It into Practice

Part IV. Epilogue

Foreword

Some years back, Tim Berners-Lee opined that we would know that the semantic web was becoming a success when people stopped asking "why?" and started asking "how?"—the same way they did with the World Wide Web many years earlier. With this book, I finally feel comfortable saying we have turned that corner. This book is about the "how"—it provides the tools a programmer needs to get going *now*!

This book's approach to the semantic web is well matched to the community that is most actively ready to start exploiting these new web technologies: programmers. More than a decade ago, researchers such as myself started playing with some of the ideas behind the semantic web, and from about 1999 to 2005, significant research funding went into the field. The "noise" from all those researchers sometimes obscured the fact that the practical technology spinning off of this research was not rocket science. In fact, that technology, which you will read about in this book, has been maturing extremely well, and it is now becoming an important component of the web developer's toolkit.

In 2000 and 2001, articles about the semantic web started to appear in the memespace of the Web. Around 2005, we started to see not just small companies in the space, but some bigger players like Oracle embracing the technology. Late in 2006, John Markoff wrote a *New York Times* article referring to "Web 3.0," and more developers started to take a serious look at the semantic web—and they liked what they saw. This developer community has helped create the tools and technologies so that, here in 2009, we're starting to see a real take-off happening. Announcements of different uses of semantic web and related technologies are appearing on an almost daily basis.

Semantic web technologies are being used by the Obama administration to provide transparency to government data, a move also being explored by many other governments around the world. Google and Yahoo! now collect and process embedded RDFa from web documents, and Microsoft recently discussed some of its semantic efforts in language-based web applications. Web 3.0 applications are attracting the sorts of user numbers that brought the early Web 2.0 apps to public attention, while a bunch of innovative startups you may not have heard of yet are exploring how to bring semantic technologies into an ever-widening range of web applications.

With all this excitement, however, has come an obvious problem. There are now a lot more people asking "how?", but since this technology is just coming into its own, there aren't many people who know how to answer the question. Where the early semantic web evangelists like me have gotten pretty good at explaining the vision to a wide range of people, including database administrators, government employees, industrialists, and academics, the questions being asked lately have been harder and harder to address. When the CTO of a Fortune 500 company asks me why he should pay attention to the technology, I can't wait to answer. However, when his developer asks me how best to find the appropriate objects for the predicates expressed in some embedded RDFa, or how the bindings of a BNode in the OPTIONAL clause of a SPARQL query work, I know that I'm soon going to be out of my depth. With the publication of this book, however, I can now point to it and say, "The answer's in there." The hole in the literature about how to make the semantic web work from the programmer's viewpoint has finally been filled.

This book also addresses another important need. Given that the top of the semantic web "layer cake" (see Chapter 11) is still in the research world, there's been a lot of confusion. On one hand, terms like "Linked Data" and "Web 3.0" are being used to describe the immediately applicable and rapidly expanding technology that is needed for web applications *today*. Meanwhile, people are also exploring the "semantic web 2.0" developments that will power the next generation. This book provides an easy way for the reader to tell the "practical now" from the pie in the sky.

Finally, I like this book for another reason: it embraces a philosophy I've often referred to as "a little Semantics goes a long way."* On the Web, a developer doesn't need to be a philosopher, an AI researcher, or a logician to understand how to make the semantic web work for him. However, figuring out just how much knowledge is enough to get going is a real challenge. In this book, Toby, Jamie, and Colin will show you "just enough RDF" (Chapter 4) and "just enough OWL" (Chapter 6) to allow you, the programmer, to get in there and start hacking.

In short, the technologies are here, the tools are ready, and this book will show you how to make it all work for you. So what are you waiting for? The future of the Web is at your fingertips.

—Jim Hendler
Albany, NY
March 2009

* *http://www.cs.rpi.edu/~hendler/LittleSemanticsWeb.html*

Preface

Like biological organisms, computers operate in complex, interconnected environments where each element of the system constrains the behavior of many others. Similar to predator-prey relationships, applications and the data they consume tend to follow co-evolutionary paths. Cumulative changes in an application eventually require modification to the data structures on which it operates. Conversely, when enhancements to a data source emerge, the structures for expressing the additional information generally force applications to change. Unfortunately, because of the significant efforts involved, this type of lock-step evolution tends to dampen enhancements in both applications and data sources.

At their core, semantic technologies decouple applications from data through the use of a simple, abstract model for knowledge representation. This model releases the mutual constraints on applications and data, allowing both to evolve independently. And by design, this degree of application-data independence promotes data portability. Any application that understands the model can consume any data source using the model. It is this level of data portability that underlies the notion of a machine-readable semantic web.

The current Web works well because we as humans are very flexible data processors. Whether the information on a web page is arranged as a table, an outline, or a multi-page narrative, we are able to extract the important information and use it to guide further knowledge discovery. However, this heterogeneity of information is indecipherable to machines, and the wide range of representations for data on the Web only compounds the problem. If the diversity of information available on the Web can be encoded by content providers into semantic data structures, any application could access and use the rich array of data we have come to rely on. In this vision, data is seamlessly woven together from disparate sources, and new knowledge is derived from the confluence. This is the vision of the semantic web.

Now, whether an application can do anything interesting with this wealth of data is where you, the developer, come into the story! Semantic technologies allow you to focus on the behavior of your applications instead of on the data processing. What does this system do when given new data sources? How can it use enhanced data models? How does the user experience improve when multiple data sources enrich one another?

Partitioning low-level data operations from knowledge utilization allows you to concentrate on what drives value in your application.

While the vision of the semantic web holds a great deal of promise, the real value of this vision is the technology that it has spawned for making data more portable and extensible. Whether you're writing a simple "mashup" or maintaining a high-performance enterprise solution, this book provides a standard, flexible approach for integrating and future-proofing systems and data.

Conventions Used in This Book

The following typographical conventions are used in this book:

Italic

> Indicates new terms, URLs, email addresses, filenames, and file extensions.

`Constant width`

> Used for program listings, as well as within paragraphs to refer to program elements such as variable or function names, databases, data types, environment variables, statements, and keywords.

 This icon signifies a tip, suggestion, or general note.

 This icon indicates a warning or caution.

Using Code Examples

This book is here to help you get your job done. In general, you may use the code in this book in your programs and documentation. You do not need to contact us for permission unless you're reproducing a significant portion of the code. For example, writing a program that uses several chunks of code from this book does not require permission. Selling or distributing a CD-ROM of examples from O'Reilly books does require permission. Answering a question by citing this book and quoting example code does not require permission. Incorporating a significant amount of example code from this book into your product's documentation does require permission.

We appreciate, but do not require, attribution. An attribution usually includes the title, author, publisher, and ISBN. For example: "*Programming the Semantic Web* by Toby Segaran, Colin Evans, and Jamie Taylor. Copyright 2009 Toby Segaran, Colin Evans, and Jamie Taylor, 978-0-596-15381-6."

If you feel your use of code examples falls outside fair use or the permission given above, feel free to contact us at *permissions@oreilly.com*.

Safari® Books Online

Safari *Books Online* When you see a Safari® Books Online icon on the cover of your favorite technology book, that means the book is available online through the O'Reilly Network Safari Bookshelf.

Safari offers a solution that's better than e-books. It's a virtual library that lets you easily search thousands of top tech books, cut and paste code samples, download chapters, and find quick answers when you need the most accurate, current information. Try it for free at *http://my.safaribooksonline.com*.

How to Contact Us

Please address comments and questions concerning this book to the publisher:

O'Reilly Media, Inc.
1005 Gravenstein Highway North
Sebastopol, CA 95472
800-998-9938 (in the United States or Canada)
707-829-0515 (international or local)
707-829-0104 (fax)

We have a web page for this book, where we list errata, examples, and any additional information. You can access this page at:

http://www.oreilly.com/catalog/9780596153816

To comment or ask technical questions about this book, send email to:

bookquestions@oreilly.com

For more information about our books, conferences, Resource Centers, and the O'Reilly Network, see our website at:

http://www.oreilly.com

The authors have established a website as a community resource for demonstrating practical approaches to semantic technology. You can access this site at:

http://www.semprog.com

## PART I

Semantic Data

## CHAPTER 1

Why Semantics?

Natural language is amazing. Without any effort you can ask a stranger how to find the nearest coffee shop; you can share your knowledge of music and martini making with your community of friends; you can go to the library, pick up a book, and learn from an author who lived hundreds of years ago. It is hard to imagine a better API for knowledge.

As a simple example, think about the following two sentences. Both are of the form "subject-verb-object," one of the simplest possible grammatical structures:

1. Colin enjoys mushrooms.
2. Mushrooms scare Jamie.

Each of these sentences represents a piece of information. The words "Jamie" and "Colin" refer to specific people, the word "mushroom" refers to a class of organisms, and the words "enjoys" and "scare" tell you the relationship between the person and the organism. Because you know from previous experience what the verbs "enjoy" and "scare" mean, and you've probably seen a mushroom before, you're able to understand the two sentences. And now that you've read them, you're equipped with new knowledge of the world. This is an example of semantics: symbols can refer to things or concepts, and sequences of symbols convey meaning. You can now use the meaning that you derived from the two sentences to answer simple questions such as "Who likes mushrooms?"

Semantics is the process of communicating enough meaning to result in an action. A sequence of symbols can be used to communicate meaning, and this communication can then affect behavior. For example, as you read this page, you are integrating the ideas expressed in these words with all that you already know. If the semantics of our writing in this book is clear, it should help you create new software, solve hard problems, and do great things.

But this book isn't about natural language; rather, it's about using semantics to represent, combine, and share knowledge between communities of machines, and how to write systems that act on that knowledge.

If you have ever written a program that used even a single variable, then you have programmed with semantics. As a programmer, you knew that this variable represented a value, and you built your program to respond to changes in the variable. Hopefully you also provided some comments in the code that explained what the variable represented and where it was used so other programmers could understand your code more easily. This relationship between the value of the variable, the meaning of the value, and the action of the program is important, but it's also implicit in the design of the system.

With a little work you can make the semantic relationships in your data explicit, and program in a way that allows the behavior of your systems to change based on the meaning of the data. With the semantics made explicit, other programs, even those not written by you, can seamlessly use your data. Similarly, when you write programs that understand semantic data, your programs can operate on datasets that you didn't anticipate when you designed your system.

Data Integration Across the Web

For applications that run on a single machine, documenting the semantics of variables in comments and documentation is adequate. The only people who need to understand the meaning of a variable are the programmers reading the source code. However, when applications participate in larger networks, the meanings of the messages they exchange need to be explicit.

Before the World Wide Web, when a user wanted to use an Internet application, he would install a tool capable of handling specific types of network messages on his machine. If a user wanted to locate users on other networks, he would install an application capable of utilizing the FINGER protocol. If a user wanted to exchange email across a network, he would install an application capable of utilizing the SMTP protocol. Each tool understood the message formats and protocols specific to its task and knew how best to display the information to the user.

Application developers would agree on the format of the messages and the behavior of applications through the circulation of RFC (Request For Comments) documents. These RFCs were written in English and made the semantics of the data contained in the messages explicit, frequently walking the reader through sample data exchanges to eliminate ambiguity. Over time, the developer community would refine the semantics of the messages to improve the capabilities of the applications. RFCs would be amended to reflect the new semantics, and application developers would update applications to make use of the new messages. Eventually users would update the applications on their machines and benefit from these new capabilities.

The emergence of the Web represented a radical change in how most people used the Internet. The Web shielded users from having to think about the applications handling the Internet messages. All you had to do was install a web browser on your machine, and any application on the Web was at your command. For developers, the Web provided a single, simple abstraction for delivering applications and made it possible for an application running in a fixed location and maintained by a stable set of developers to service all Internet users.

Underlying the Web is a set of messages that developers of web infrastructure have agreed to treat in a standard manner. It is well understood that when a web server speaking HTTP receives a GET request, it should send back data corresponding to the path portion of the request message. The semantics of these messages have been thoroughly defined by standards committees and documented in RFCs and W3C recommendations. This standardized infrastructure allows web application developers to operate behind a facade that separates them from the details of how application data is transmitted between machines, and focus on how their applications appear to users. Web application developers no longer need to coordinate with other developers about message formats or how applications should behave in the presence of certain data.

While this facade has facilitated an explosion in applications available to users, the decoupling of data transmission from applications has caused data to become compartmentalized into stovepipe systems, hidden behind web interfaces. The web facade has, in effect, prevented much of the data fueling web applications from being shared and integrated into other Internet applications.

Applications that combine data in new ways and allow users to make connections and understand relationships that were previously hidden are very powerful and compelling. These applications can be as simple as plotting crime statistics on a map or as informative as showing which cuisines are available within walking distance of a film that you want to watch. But currently the process to build these applications is highly specialized and idiosyncratic, with each application using hand-tuned and ad-hoc techniques for harvesting and integrating information due to the hidden nature of data on the Web.

This book introduces repeatable approaches to these data integration problems through the use of simple mechanisms that explicitly expose the semantics of data. These mechanisms provide standardized ways for data to be published and combined, allowing developers to focus on building data-rich applications rather than getting stuck on problems of obtaining and integrating data.

Traditional Data-Modeling Methods

There are many ways to model data, some of them very well researched and mature. In this book we explore new ways to model data, but we're certainly not trying to convince you that the old ways are wrong. There are many ways to think about data, and it is important to have a wide range of tools available so you can pick the best one for the task at hand.

In this section, we'll look at common methods that you've likely encountered and consider their strengths and weaknesses when it comes to integrating data across the Web and in the face of quickly changing requirements.

Tabular Data

The simplest kind of dataset, and one that almost everyone is familiar with, is tabular data. Tabular data is any data kept in a table, such as an Excel spreadsheet or an HTML table. Tabular data has the advantage of being very simple to read and manipulate. Consider the restaurant data shown in Table 1-1.

Table 1-1. A table of restaurants

Restaurant	Address	Cuisine	Price	Open
Deli Llama	Peachtree Rd	Deli	$	Mon, Tue, Wed, Thu, Fri
Peking Inn	Lake St	Chinese	$$$	Thu, Fri, Sat
Thai Tanic	Branch Dr	Thai	$$	Tue, Wed, Thu, Fri, Sat, Sun
Lord of the Fries	Flower Ave	Fast Food	$$	Tue, Wed, Thu, Fri, Sat, Sun
Marquis de Salade	Main St	French	$$$	Thu, Fri, Sat
Wok This Way	Second St	Chinese	$	Mon, Tue, Wed, Thu, Fri, Sat, Sun
Luna Sea	Autumn Dr	Seafood	$$$	Tue, Thu, Fri, Sat
Pita Pan	Thunder Rd	Middle Eastern	$$	Mon, Tue, Wed, Thu, Fri, Sat, Sun
Award Weiners	Dorfold Mews	Fast Food	$	Mon, Tue, Wed, Thu, Fri, Sat
Lettuce Eat	Rustic Parkway	Deli	$$	Mon, Tue, Wed, Thu, Fri

Data kept in a table is generally easy to display, sort, print, and edit. In fact, you might not even think of data in an Excel spreadsheet as "modeled" at all, but the placement of the data in rows and columns gives each piece a particular meaning. Unlike the modeling methods we'll see later, there's not really much variation in the ways you might look at tabular data. It's often said that most "databases" used in business settings are simply spreadsheets.

It's interesting to note that there are *semantics* in a data table or spreadsheet: the row and column in which you choose to put the data—for example, a restaurant's cuisine—explains what the name means to a person reading the data. The fact that *Chinese* is in the row *Peking Inn* and in the column *Cuisine* tells us immediately that "Peking Inn serves Chinese food." You know this because you understand what restaurants and cuisines are and because you've previously learned how to read a table. This may seem like a trivial point, but it's important to keep in mind as we explore different ways to think about data.

Data stored this way has obvious limitations. Consider the last column, *Open*. You can see that we've crammed a list of days of the week into a single column. This is fine if all we're planning to do is read the table, but it breaks down if we want to add more information such as the open hours or nightly specials. In theory, it's possible to add this information in parentheses after the days, as shown in Table 1-2.

Table 1-2. Forcing too much data into a spreadsheet

Restaurant	Address	Cuisine	Price	Open
Deli Llama	Peachtree Rd	Deli	$	Mon (11a–4p), Tue (11–4), Wed (11–4), Thu (11–7), Fri (11–8)
Peking Inn	Lake St	Chinese	$$$	Thu (5p–10p), Fri (5–11), Sat (5–11)

However, we can't use this data in a spreadsheet program to find the restaurants that will be open late on Friday night. Sorting on the columns simply doesn't capture the deeper meaning of the text we've entered. The program doesn't understand that we've used individual fields in the *Open* column to store multiple distinct information values.

The problems with spreadsheets are compounded when we have multiple spreadsheets that make reference to the same data. For instance, if we have a spreadsheet of our friends' reviews of the restaurants listed earlier, there would be no easy way to search across both documents to find restaurants near our homes that our friends recommend. Although Excel experts can often use macros and lookup tables to get the spreadsheet to approximate this desired behavior, the models are rigid, limited, and usually not changeable by other users.

The need for a more sophisticated way to model data becomes obvious very quickly.

Relational Data

It's almost impossible for a programmer to be unfamiliar with relational databases, which are used in all kinds of applications in every industry. Products like Oracle DB, MySQL, and PostgreSQL are very mature and are the result of years of research and optimization. Relational databases are very fast and powerful tools for storing large sets of data where the data model is well understood and the usage patterns are fairly predictable.

Essentially, a relational database allows multiple tables to be joined in a standardized way. To store our restaurant data, we might define a schema like the one shown in Figure 1-1. This allows our restaurant data to be represented in a more useful and flexible way, as shown in Figure 1-2.

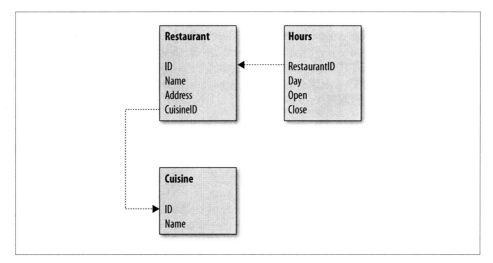

Figure 1-1. Simple restaurant schema

Restaurant				
ID	Name	Address	Price	CuisineID
1	Deli Llama	Peachtree Rd	$	1
2	Peking Inn	Lake St	$$$	2

Cuisine	
ID	Name
1	Deli
2	Chinese
3	Thai
4	Fast Food

Hours			
RestID	Day	Open	Close
1	Mon	11	16
1	Tue	11	16
1	Wed	11	16
1	Thu	11	19
1	Fri	11	20
2	Thu	5	22
2	Fri	5	23
2	Sat	5	23

Figure 1-2. Relational restaurant data

Now, instead of sorting or filtering on a single column, we can do more sophisticated queries. A query to find all the restaurants that will be open at 10 p.m. on a Friday can be expressed using SQL like this:

```
SELECT Restaurant.Name, Cuisine.Name, Hours.Open, Hours.Close
FROM Restaurant, Cuisine, Hours
WHERE Restaurant.CuisineID=Cuisine.ID
AND Restaurant.ID=Hours.RestaurantID
AND Hours.Day="Fri"
AND Hours.Open<22
AND Hours.Close>22
```

which gives a result like this:

```
Restaurant.Name   |   Cuisine.Name   |   Hours.Open   |   Hours.Close   |
------------------------------------------------------------------------
Peking Inn        |   Chinese        |   17           |   23            |
```

Notice that in our relational data model, the semantics of the data have been made more explicit. The meanings of the values are actually described by the schema: someone looking at the tables can immediately see that there are several types of entities modeled—a type called "restaurant" and a type called "days"—and that they have specific relationships between them. Furthermore, even though the database doesn't really know what a "restaurant" is, it can respond to requests to see all the restaurants with given properties. Each datum is labeled with what it means by virtue of the table and column that it's in.

Evolving and Refactoring Schemas

The previous section mentioned that relational databases are great for datasets where the data model is understood up front and there is some understanding of how the data will be used. Many applications, such as product catalogs or contact lists, lend themselves well to relational schemas, since there are generally a fixed set of fields and a set of fairly typical usage patterns.

However, as we've been discussing, data integration across the Web is characterized by rapidly changing types of data, and programmers can never quite know what will be available and how people might want to use it. As a simple example, let's assume we have our restaurant database up and running, and then we receive a new database of bars with additional information not in our restaurant schema, as shown in Table 1-3.

Table 1-3. A new dataset of bars

Bar	Address	DJ	Specialty drink
The Bitter End	14th Ave	No	Beer
Peking Inn	Lake St	No	Scorpion Bowl
Hammer Time	Wildcat Dr	Yes	Hennessey
Marquis de Salade	Main St	Yes	Martini

Of course, many restaurants also have bars, and as it gets later in the evening, they may stop serving food entirely and only serve drinks. The table of bars in this case shows that, in addition to being a French restaurant, Marquis de Salade is also a bar with a DJ. The table also shows specialty drinks, which gives us additional information about Marquis. As of right now, these databases are separate, but it's certainly possible that someone might want to query across them—for example, to find a place to get a French meal and stay later for martinis.

So how do we update our database so that it supports the new bar data? Well, we could just link the tables with another table, which has the upside of not forcing us to change the existing structure. Figure 1-3 shows a database structure with an additional table, RB_Link, that links the existing tables, telling you when a restaurant and a bar are actually the same place.

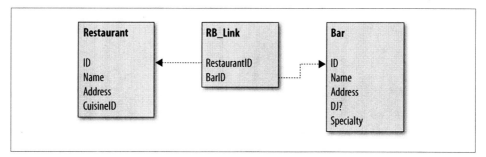

Figure 1-3. Linking bars to the existing schema

This works, and certainly makes our query possible, but it introduces a problem: there are now two names and addresses in our database for establishments that are both bars and restaurants, and just a link telling us that they're the same place. If you want to query by address, you need to look at both tables. Also, the type of food served is attached to the restaurant type, but not to its bar type. Adding and updating data is much more complicated.

Perhaps a more accurate way to model this would be to have a Venue table with bar and restaurant types separated out, like the one shown in Figure 1-4.

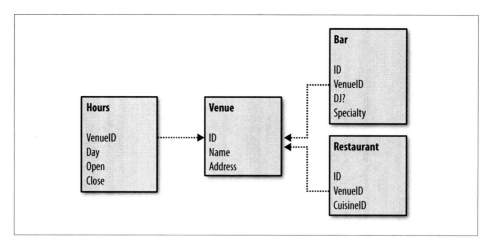

Figure 1-4. Normalized schema that separates a venue from its purposes

This seems to solve our issues, but remember that all the existing data is still in our old data model and needs to be transformed to the new data model. This process is called *schema migration* and is often a huge headache. Not only does the data have to be migrated, but all the queries and dependent code that was written assuming a certain table structure have to be changed as well. If we have built a restaurant website on top of our old schema, then we need to figure out how to update all of our existing code, queries, and the database without imposing significant downtime on the website. A whole discipline of software engineering has emerged to deal with these issues, using techniques like stored database procedures and object-relational mapping (ORM) layers to try to decouple the underlying schema from the business-logic layer and lowering the cost of schema changes. These techniques are useful, but they impose their own complexities and problems as well.

It's easy to imagine that, as our restaurant application matures, these venues could also have all kinds of other uses such as a live music hall or a rental space for events. When dealing with data integration across the entire Web, or even in smaller environments that are constantly facing new datasets, migrating the schema each time a new type of data is encountered is simply not tractable. Too often, people have to resort to manual lookups, overly convoluted linked spreadsheets, or just setting the data aside until they can decide what to do with it.

Very Complicated Schemas

In addition to having to migrate as the data evolves, another problem one runs into with relational databases is that the schemas can get incredibly complicated when dealing with many different kinds of data. For example, Figure 1-5 shows a small section of the schema for a Customer Relationship Management (CRM) product.

A CRM system is used to store information about customer leads and relationships with current customers. This is obviously a big application, but to put things in perspective, it is a very small piece of what is required to run a business. An ERP (Enterprise Resource Planning) application, such as SAP, can cover many more of the data needs of a large business. However, the schemas for ERP applications are so inaccessible that there is a whole industry of consulting companies that exclusively deal with them.

The complexity is barely manageable in well-understood industry domains like CRM and ERP, but it becomes even worse in rapidly evolving fields such as biotechnology and international development. Instead of a few long lists of well-characterized data, we instead have hundreds or thousands of datasets, all of which talk about very different things. Trying to normalize these to a single schema is a labor-intensive and painful process.

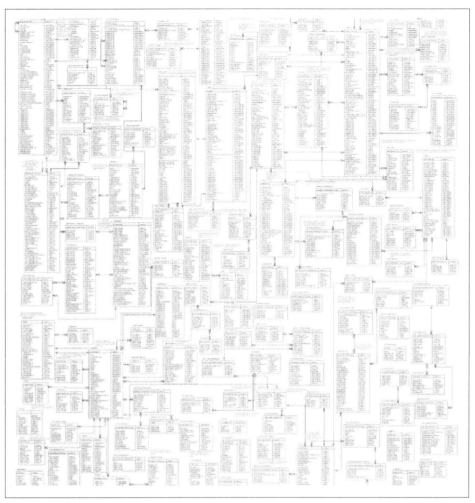

Figure 1-5. Example of a big schema

Getting It Right the First Time

This brings us to the question of whether it's possible to define a schema so that it's flexible enough to handle a wide variety of ever-changing types of data, while still maintaining a certain level of readability. Maybe the schema could be designed to be open to new venue purposes and offer custom fields for them, something like Figure 1-6. The schema has lost the concepts of bars and restaurants entirely, now containing just a list of venues and custom properties for them.

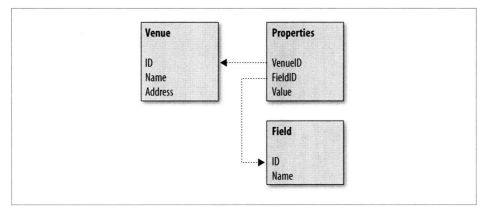

Figure 1-6. Venue schema with completely custom properties

This is not usually recommended, as it gets rid of a lot of the normalization that was possible before and will likely degrade the performance of the database. However, it allows us to express the data in a way that allows for new venue purposes to come along, an example of which is shown in Figure 1-7. Notice how the Properties table contains all the custom information for each of the venues.

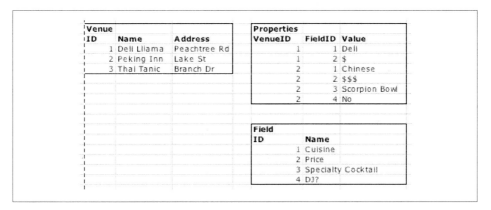

Figure 1-7. Venue data in more flexible form

This means that the application can be extended to include, for example, concert venues. Maybe we're visiting a city and looking for a place to stay close to cheap food and cool concert venues. We could create new fields in the Field table, and then add custom properties to any of the existing venues. Figure 1-8 shows an example where we've added the information that Thai Tanic has live jazz music. There are two extra fields, "Live Music?" and "Music Genre", and two more rows in the Properties table.

Venue		
ID	Name	Address
1	Deli Llama	Peachtree Rd
2	Peking Inn	Lake St
3	Thai Tanic	Branch Dr

Properties		
VenueID	FieldID	Value
1	1	Deli
1	2	$
2	1	Chinese
2	2	$$$
2	3	Scorpion Bowl
2	4	No
3	5	Yes
3	6	Jazz

Field	
ID	Name
1	Cuisine
2	Price
3	Specialty Cocktail
4	DJ?
5	Live Music
6	Music Genre

Figure 1-8. Adding a concert venue without changing the schema

This type of key/value schema extension is nothing new, and many people stumble into this kind of representation when they have sparse relationships to represent. In fact, many "customizable" data applications such as Saleforce.com represent data this way internally. However, because this type of representation turns the database tables "on their sides," database performance frequently suffers, and therefore it is generally not considered a good idea (i.e., best practice). We've also lost a lot of the normalization we were able to do before, because we've flattened everything to key/value pairs.

Semantic Relationships

Even though it might not be considered a best practice, let's continue with this progression and see what happens. Why not move all the relationships expressed in standard table rows into this parameterized key/value format? From this perspective, the venue name and address are just properties of the venue, so let's move the columns in the Venue table into key/value rows in the Properties table. Figure 1-9 shows what this might look like.

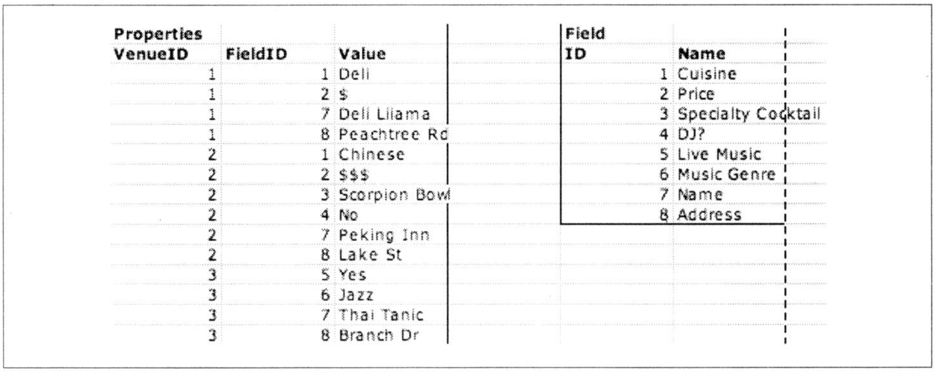

Figure 1-9. Parameterized venues

That's interesting, but the relationship between the Properties table and the Field table is still only known through the knowledge trapped in the logic of our join query. Let's make that knowledge explicit by preforming the join and displaying the result set in the same parameterized way (Table 1-4).

Table 1-4. Fully parameterized venues

Properties

VenueID	Field	Value
1	Cuisine	Deli
1	Price	$
1	Name	Deli Llama
1	Address	Peachtree Rd
2	Cuisine	Chinese
2	Price	$$$
2	Specialty Cocktail	Scorpion Bowl
2	DJ?	No
2	Name	Peking Inn
2	Address	Lake St
3	Live Music?	Yes
3	Music Genre	Jazz
3	Name	Thai Tanic
3	Address	Branch Dr

Now each datum is described alongside the property that defines it. In doing this, we've taken the semantic relationships that previously were inferred from the table and column and made them data in the table. This is the essence of semantic data modeling: flexible schemas where the relationships are described *by the data itself*. In the remainder of this book, you'll see how you can move *all* of the semantics into the data. We'll show you how to represent data in this manner, and we'll introduce tools especially designed for storing, visualizing, and querying semantic data.

Metadata Is Data

One of the challenges of using someone else's relational data is understanding how the various tables relate to one another. This information—the data about the data representation—is often called metadata and represents knowledge about how the data can be used. This knowledge is generally represented explicitly in the data definition through foreign key relationships, or implicitly in the logic of the queries. Too frequently, data is archived, published, or shared without this critical metadata. While rediscovering these relationships can be an exciting exercise for the user, schemas need not become very large before metadata recovery becomes nearly impossible.

In our earlier example, parameterizing the venue data made the model extremely flexible. When we learn of a new characteristic for a venue, we simply need to add a new row to the table, even if we've never seen that characteristic before. Parameterizing the data also made it trivial to use. You need very little knowledge about the organization of the data to make use of it. Once you know that rows sharing identical VenueIDs relate to one another, you can learn everything there is to know about a venue by selecting all rows with the same VenueID. From this perspective, we can think of the parameterized venue data as "self-describing data." The metadata of the relational schema, describing which columns go together to describe a single entity, has become part of the data itself.

Building for the Unexpected

By packing the data and metadata together in a single representation, we have not only made the schema future-proof, we have also isolated our applications from "knowing" too much about the form of the data. The only thing our application needs to know about the data is that a venue will have an ID in the first column, the properties of the venue appear in the second column, and the third column represents the value of each property. When we add a totally new property to a venue, the application can either choose to ignore the property or handle it in a standard manner.

Because our data is represented in a flexible model, it is easy for someone else to integrate information about espresso machine locations, allowing our application to cover not only restaurants and bars, but also coffee shops, book stores, and gas stations (at least in the greater Seattle area). A well-designed application should be able to seamlessly integrate new semantic data, and semantic datasets should be able to work with a wide variety of applications.

Many content and image creation tools now support XMP (Extensible Metadata Platform) data for tracking information about the author and licensing of creative content. The XMP standard, developed by Adobe Systems, provides a standard set of schemas and allows users to extend the data model in exactly the way we extended the venue data. By using a self-describing model, the tools used to inspect content for XMP data need not change, even if the types of content change significantly in the future. Since image creation tools are fundamentally for creative expression, it is essential that users not be limited to a fixed set of descriptive fields.

"Perpetual Beta"

It's clear that the Web changed the economics of application development. The web facade greatly reduced coordination costs by cutting applications free from the complexity of managing low-level network data messages. With a single application capable of servicing all the users on the Internet, the software development deadlines imposed by manufacturing lead time and channel distribution are quaint memories for most of us. Applications are now free to improve on a continuous and independent basis. Development cycles that update application code on a monthly, weekly, or even daily basis are no longer considered unusual. The phrase "perpetual beta" reflects this sentiment that software is never "frozen" on the Web. As applications continually improve, continuous release processes allow users to instantaneously benefit.

Compressed release cycles are a part of staying competitive at large portal sites. For example, Yahoo! has a wide variety of media sites covering topics such as health, kids, travel, and movies. Content is continually changing as news breaks, editorial processes complete, and users annotate information. In an effort to reduce the time necessary to produce specialized websites and enable new types of personalization and search, Yahoo! has begun to add semantic metadata to their content using extensible schemas not unlike the examples developed here. As data and metadata become one, new applications can add their own annotations without modification to the underlying schema. This freedom to extend the existing metadata enables constantly evolving features without affecting legacy applications, and it allows one application to benefit from the information provided by another.

This shift to continually improving and evolving applications has been accompanied by a greater interest in what were previously considered "scripting" languages such as Python, Perl, and Ruby. The ease of getting something up and running with minimal upfront design and the ease of quick iterations to add new features gives these languages an advantage over heavier static languages that were designed for more traditional approaches to software engineering. However, most frameworks that use these languages still rely on relational databases for storage, and thus still require upfront data modeling and commitment to a schema that may not support the new data sources that future application features require.

So, while perpetual beta is a great benefits to users, rapid development cycles can be a challenge for data management. As new application features evolve, data schemas are frequently forced to evolve as well. As we will see throughout the remainder of this book, flexible semantic data structures and the application patterns that work with them are well designed for life in a world of perpetual beta.

## CHAPTER 2

Expressing Meaning

In the previous chapter we showed you a simple yet flexible data structure for describing restaurants, bars, and music venues. In this chapter we will develop some code to efficiently handle these types of data structures. But before we start working on the code, let's see if we can make our data structure a bit more robust.

In its current form, our "fully parameterized venue" table allows us to represent arbitrary facts about food and music venues. But why limit the table to describing just these kinds of items? There is nothing specific about the form of the table that restricts it to food and music venues, and we should be able to represent facts about other entities in this same three-column format.

In fact, this three-column format is known as a *triple*, and it forms the fundamental building block of semantic representations. Each triple is composed of a subject, a predicate, and an object. You can think of triples as representing simple linguistic statements, with each element corresponding to a piece of grammar that would be used to diagram a short sentence (see Figure 2-1).

Figure 2-1. Sentence diagram showing a subject-predicate-object relationship

Generally, the subject in a triple corresponds to an entity—a "thing" for which we have a conceptual class. People, places, and other concrete objects are entities, as are less concrete things like periods of time and ideas. Predicates are a property of the entity to which they are attached. A person's name or birth date or a business's stock symbol or mailing address are all examples of predicates. Objects fall into two classes: entities that can be the subject in other triples, and literal values such as strings or numbers.

Multiple triples can be tied together by using the same subjects and objects in different triples, and as we assemble these chains of relationships, they form a directed graph.

Directed graphs are well-known data structures in computer science and mathematics, and we'll be using them to represent our data.

Let's apply our graph model to our venue data by relaxing the meaning of the first column and asserting that IDs can represent any entity. We can then add neighborhood information to the same table as our restaurant data (see Table 2-1).

Table 2-1. Extending the Venue table to include neighborhoods

Subject	Predicate	Object
S1	Cuisine	"Deli"
S1	Price	"$"
S1	Name	"Deli Llama"
S1	Address	"Peachtree Rd"
S2	Cuisine	"Chinese"
S2	Price	"$$$"
S2	Specialty Cocktail	"Scorpion Bowl"
S2	DJ?	"No"
S2	Name	"Peking Inn"
S2	Address	"Lake St"
S3	Live Music?	"Yes"
S3	Music Genre	"Jazz"
S3	Name	"Thai Tanic"
S3	Address	"Branch Dr"
S4	Name	"North Beach"
S4	Contained-by	"San Francisco"
S5	Name	"SOMA"
S5	Contained-by	"San Francisco"
S6	Name	"Gourmet Ghetto"
S6	Contained-by	"Berkeley"

Now we have venues and neighborhoods represented using the same model, but nothing connects them. Since objects in one triple can be subjects in another triple, we can add assertions that specify which neighborhood each venue is in (see Table 2-2).

Table 2-2. The triples that connect venues to neighborhoods

Subject	Predicate	Object
S1	Has Location	S4
S2	Has Location	S6
S3	Has Location	S5

Figure 2-2 is a diagram of some of our triples structured as a graph, with subjects and objects as nodes and predicates as directed arcs.

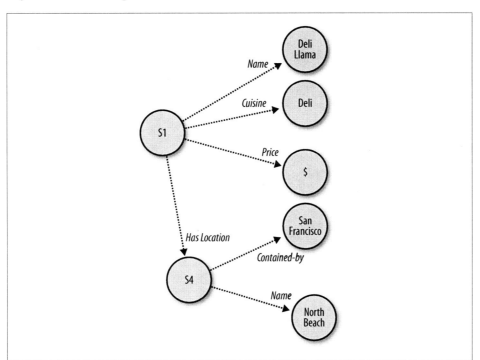

Figure 2-2. A graph of triples showing information about a restaurant

Now, by following the chain of assertions, we can determine that it is possible to eat cheaply in San Francisco. You just need to know where to look.

An Example: Movie Data

We can use this triple model to build a simple representation of a movie. Let's start by representing the title of the movie *Blade Runner* with the triple (`blade_runner` name "`Blade Runner`"). You can think of this triple as an arc representing the predicate called name, connecting the subject `blade_runner` to an object, in this case a string, representing the value "Blade Runner" (see Figure 2-3).

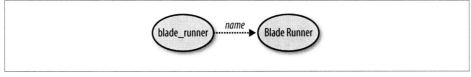

Figure 2-3. A triple describing the title of the movie Blade Runner

Now let's add the release date of the film. This can be done with another triple (`blade_runner release_date "June 25, 1982"`). We use the same ID `blade_runner`, which indicates that we're referring to the same subject when making these statements. It is by using the same IDs in subjects and objects that a graph is built—otherwise, we would have a bunch of disconnected facts and no way of knowing that they concern the same entities.

Next, we want to assert that Ridley Scott directed the movie. The simplest way to do this would be to add the triple (`blade_runner directed_by "Ridley Scott"`). There is a problem with this approach, though—we haven't assigned Ridley Scott an ID, so he can't be the source of new assertions, and we can't connect him to other movies he has directed. Additionally, if there happen to be other people named "Ridley Scott", we won't be able to distinguish them by name alone.

Ridley Scott is a person and a director, among other things, and that definitely qualifies as an entity. If we give him the ID `ridley_scott`, we can assert some facts about him: (`ridley_scott name "Ridley Scott"`), and (`blade_runner directed_by ridley_scott`). Notice that we reused the `name` property from earlier. Both entities, "Blade Runner" and "Ridley Scott", have names, so it makes sense to reuse the `name` property as long as it is consistent with other uses. Notice also that we asserted a triple that connected two entities, instead of just recording a literal value. See Figure 2-4.

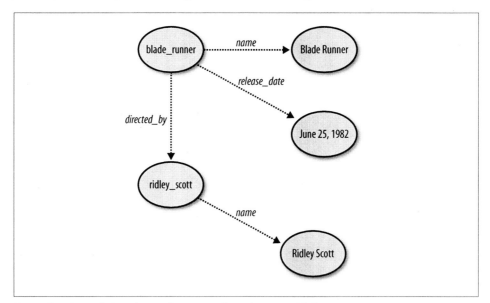

Figure 2-4. A graph describing information about the movie Blade Runner

Building a Simple Triplestore

In this section we will build a simple, cross-indexed triplestore. Since there are many excellent semantic toolkits available (which we will explore in more detail in later chapters), there is really no need to write a triplestore yourself. But by working through the scaled-down code in this section, you will gain a better understanding of how these systems work.

Our system will use a common triplestore design: cross-indexing the subject, predicate, and object in three different permutations so that all triple queries can be answered through lookups. All the code in this section is available to download at *http://semprog .com/psw/chapter2/simpletriple.py*. You can either download the code and just read the section to understand what it's doing, or you can work through the tutorial to create the same file.

The examples in this section and throughout the book are in Python. We chose to use Python because it's a very simple language to read and understand, it's concise enough to fit easily into short code blocks, and it has a number of useful toolkits for semantic web programming. The code itself is fairly simple, so programmers of other languages should be able to translate the examples into the language of their choice.

Indexes

To begin with, create a file called *simplegraph.py*. The first thing we'll do is create a class that will be our triplestore and add an initialization method that creates the three indexes:

```python
class SimpleGraph:
    def __init__(self):
        self._spo = {}
        self._pos = {}
        self._osp = {}
```

Each of the three indexes holds a different permutation of each triple that is stored in the graph. The name of the index indicates the ordering of the terms in the index (i.e., the pos index stores the **p**redicate, then the **o**bject, and then the **s**ubject, in that order). The index is structured using a dictionary containing dictionaries that in turn contain sets, with the first dictionary keyed off of the first term, the second dictionary keyed off of the second term, and the set containing the third terms for the index. For example, the pos index could be instantiated with a new triple like so:

```python
self._pos = {predicate:{object:set([subject])}}
```

A query for all triples with a specific predicate and object could be answered like so:

```python
for subject in self._pos[predicate][object]: yield (subject, predicate, object)
```

Each triple is represented in each index using a different permutation, and this allows any query across the triples to be answered simply by iterating over a single index.

The add and remove Methods

The add method permutes the subject, predicate, and object to match the ordering of each index:

```
def add(self, (sub, pred, obj)):
    self._addToIndex(self._spo, sub, pred, obj)
    self._addToIndex(self._pos, pred, obj, sub)
    self._addToIndex(self._osp, obj, sub, pred)
```

The _addToIndex method adds the terms to the index, creating a dictionary and set if the terms are not already in the index:

```
def _addToIndex(self, index, a, b, c):
    if a not in index: index[a] = {b:set([c])}
    else:
        if b not in index[a]: index[a][b] = set([c])
        else: index[a][b].add(c)
```

The remove method finds all triples that match a pattern, permutes them, and removes them from each index:

```
def remove(self, (sub, pred, obj)):
    triples = list(self.triples((sub, pred, obj)))
    for (delSub, delPred, delObj) in triples:
        self._removeFromIndex(self._spo, delSub, delPred, delObj)
        self._removeFromIndex(self._pos, delPred, delObj, delSub)
        self._removeFromIndex(self._osp, delObj, delSub, delPred)
```

The _removeFromIndex walks down the index, cleaning up empty intermediate dictionaries and sets while removing the terms of the triple:

```
def _removeFromIndex(self, index, a, b, c):
    try:
        bs = index[a]
        cset = bs[b]
        cset.remove(c)
        if len(cset) == 0: del bs[b]
        if len(bs) == 0: del index[a]
    # KeyErrors occur if a term was missing, which means that it wasn't a
    # valid delete:
    except KeyError:
        pass
```

Finally, we'll add methods for loading and saving the triples in the graph to comma-separated files. Make sure to import the csv module at the top of your file:

```
def load(self, filename):
    f = open(filename, "rb")
    reader = csv.reader(f)
    for sub, pred, obj in reader:
        sub = unicode(sub, "UTF-8")
        pred = unicode(pred, "UTF-8")
        obj = unicode(obj, "UTF-8")
        self.add((sub, pred, obj))
    f.close()
```

```
def save(self, filename):
    f = open(filename, "wb")
    writer = csv.writer(f)
    for sub, pred, obj in self.triples((None, None, None)):
        writer.writerow([sub.encode("UTF-8"), pred.encode("UTF-8"), \
            obj.encode("UTF-8")])
    f.close()
```

Querying

The basic query method takes a (subject, predicate, object) pattern and returns all
triples that match the pattern. Terms in the triple that are set to None are treated as
wildcards. The triples method determines which index to use based on which terms
of the triple are wildcarded, and then iterates over the appropriate index, yielding triples
that match the pattern:

```
def triples(self, (sub, pred, obj)):
    # check which terms are present in order to use the correct index:
    try:
        if sub != None:
            if pred != None:
                # sub pred obj
                if obj != None:
                    if obj in self._spo[sub][pred]:
                        yield (sub, pred, obj)
                # sub pred None
                else:
                    for retObj in self._spo[sub][pred]:
                        yield (sub, pred, retObj)
            else:
                # sub None obj
                if obj != None:
                    for retPred in self._osp[obj][sub]:
                        yield (sub, retPred, obj)
                # sub None None
                else:
                    for retPred, objSet in self._spo[sub].items():
                        for retObj in objSet:
                            yield (sub, retPred, retObj)
        else:
            if pred != None:
                # None pred obj
                if obj != None:
                    for retSub in self._pos[pred][obj]:
                        yield (retSub, pred, obj)
                # None pred None
                else:
                    for retObj, subSet in self._pos[pred].items():
                        for retSub in subSet:
                            yield (retSub, pred, retObj)
            else:
                # None None obj
                if obj != None:
```

```
            for retSub, predSet in self._osp[obj].items():
                for retPred in predSet:
                    yield (retSub, retPred, obj)
        # None None None
        else:
            for retSub, predSet in self._spo.items():
                for retPred, objSet in predSet.items():
                    for retObj in objSet:
                        yield (retSub, retPred, retObj)
    # KeyErrors occur if a query term wasn't in the index,
    # so we yield nothing:
    except KeyError:
        pass
```

Now, we'll add a convenience method for querying a single value of a single triple:

```
def value(self, sub=None, pred=None, obj=None):
    for retSub, retPred, retObj in self.triples((sub, pred, obj)):
        if sub is None: return retSub
        if pred is None: return retPred
        if obj is None: return retObj
        break
    return None
```

That's all you need for a basic in-memory triplestore. Although you'll see more sophisticated implementations throughout this book, this code is sufficient for storing and querying all kinds of semantic information. Because of the indexing, the performance will be perfectly acceptable for tens of thousands of triples.

Launch a Python prompt to try it out:

```
>>> from simplegraph import SimpleGraph
>>> movie_graph=SimpleGraph()
>>> movie_graph.add(('blade_runner','name','Blade Runner'))
>>> movie_graph.add(('blade_runner','directed_by','ridley_scott'))
>>> movie_graph.add(('ridley_scott','name','Ridley Scott'))
>>> list(movie_graph.triples(('blade_runner','directed_by',None)))
[('blade_runner', 'directed_by', 'ridley_scott')]
>>> list(movie_graph.triples((None,'name',None)))
[('ridley_scott', 'name', 'Ridley Scott'), ('blade_runner', 'name', 'Blade Runner')]
>>> movie_graph.value('blade_runner','directed_by',None)
ridley_scott
```

Merging Graphs

One of the marvelous properties of using graphs to model information is that if you have two separate graphs with a consistent system of identifiers for subjects and objects, you can merge the two graphs with no effort. This is because nodes and relationships in graphs are first-class entities, and each triple can stand on its own as a piece of meaningful data. Additionally, if a triple is in both graphs, the two triples merge together transparently, because they are identical. Figures 2-5 and 2-6 illustrate the ease of merging arbitrary datasets.

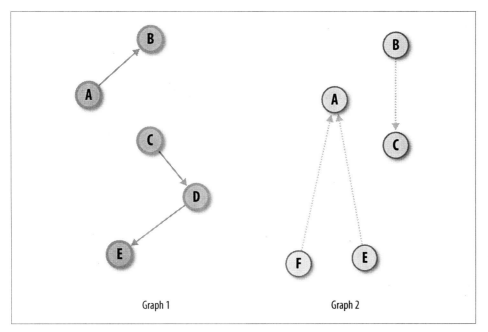

Figure 2-5. *Two separate graphs that share some identifiers*

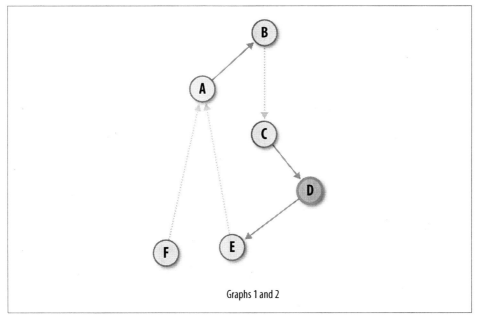

Figure 2-6. *The merged graph that is the union of triples*

In the case of our simple graph, this example will merge two graphs into a single third graph:

```
>>> graph1 = SimpleGraph()
>>> graph2 = SimpleGraph()
... load data into the graphs ...
>>> mergegraph = SimpleGraph()
>>> for sub, pred, obj in graph1:
...     mergegraph.triples((None, None, None)).add((sub, pred, obj))
>>> for sub, pred, obj in graph2:
...     mergegraph.triples((None, None, None)).add((sub, pred, obj))
```

Adding and Querying Movie Data

Now we're going to load a large set of movies, actors, and directors. The *movies.csv* file available at *http://semprog.com/psw/chapter2/movies.csv* contains over 20,000 movies and is taken from Freebase.com. The predicates that we'll be using are name, directed_by for directors, and starring for actors. The IDs for all of the entities are the internal IDs used at Freebase.com. Here's how we load it into a graph:

```
>>> import simplegraph
>>> graph = simplegraph.SimpleGraph()
>>> graph.load("movies.csv")
```

Next, we'll find the names of all the actors in the movie *Blade Runner*. We do this by first finding the ID for the movie named "Blade Runner", then finding the IDs of all the actors in the movie, and finally looking up the names of those actors:

```
>>> bladerunnerId = graph.value(None, "name", "Blade Runner")
>>> print bladerunnerId
/en/blade_runner
>>> bladerunnerActorIds = [actorId for _, _, actorId in \
... graph.triples((bladerunnerId, "starring", None))]
>>> print bladerunnerActorIds
[u'/en/edward_james_olmos', u'/en/william_sanderson', u'/en/joanna_cassidy',
u'/en/harrison_ford', u'/en/rutger_hauer', u'/en/daryl_hannah', ...
>>> [graph.value(actorId, "name", None) for actorId in bladerunnerActorIds]
[u'Edward James Olmos',u'William Sanderson', u'Joanna Cassidy', u'Harrison Ford',
u'Rutger Hauer', u'Daryl Hannah', ...
```

Next, we'll explore what other movies Harrison Ford has been in besides *Blade Runner*:

```
>>> harrisonfordId = graph.value(None, "name", "Harrison Ford")
>>> [graph.value(movieId, "name", None) for movieId, _, _ in \
... graph.triples((None, "starring", harrisonfordId))]
[u'Star Wars Episode IV: A New Hope', u'American Graffiti',
u'The Fugitive', u'The Conversation', u'Clear and Present Danger',...
```

Using Python set intersection, we can find all of the movies in which Harrison Ford has acted that were directed by Steven Spielberg:

```
>>> spielbergId = graph.value(None, "name", "Steven Spielberg")
>>> spielbergMovieIds = set([movieId for movieId, _, _ in \
... graph.triples((None, "directed_by", spielbergId))])
```

```
>>> harrisonfordId = graph.value(None, "name", "Harrison Ford")
>>> harrisonfordMovieIds = set([movieId for movieId, _, _ in \
... graph.triples((None, "starring", harrisonfordId))])
>>> [graph.value(movieId, "name", None) for movieId in \
... spielbergMovieIds.intersection(harrisonfordMovieIds)]
[u'Raiders of the Lost Ark', u'Indiana Jones and the Kingdom of the Crystal Skull',
 u'Indiana Jones and the Last Crusade', u'Indiana Jones and the Temple of Doom']
```

It's a little tedious to write code just to do queries like that, so in the next chapter we'll show you how to build a much more sophisticated query language that can filter and retrieve more complicated queries. In the meantime, let's look at a few more graph examples.

Other Examples

Now that you've learned how to represent data as a graph, and worked through an example with movie data, we'll look at some other types of data and see how they can also be represented as graphs. This section aims to show that graph representations can be used for a wide variety of purposes. We'll specifically take you through examples in which the different kinds of information could easily grow.

We'll look at data about places, celebrities, and businesses. In each case, we'll explore ways to represent the data in a graph and provide some data for you to download. All these triples were generated from Freebase.com.

Places

Places are particularly interesting because there is so much data about cities and countries available from various sources. Places also provide context for many things, such as news stories or biographical information, so it's easy to imagine wanting to link other datasets into comprehensive information about locations. Places can be difficult to model, however, in part because of the wide variety of types of data available, and also because there's no clear way to define them—a city's name can refer to its metro area or just to its limits, and concepts like counties, states, parishes, neighborhoods, and provinces vary throughout the world.

Figure 2-7 shows a graph centered around "San Francisco". You can see that San Francisco is in California, and California is in the United States. By structuring the places as a graphical hierarchy we avoid the complications of defining a city as being in a state, which is true in some places but not in others. We also have the option to add more information, such as neighborhoods, to the hierarchy if it becomes available. The figure shows various information about San Francisco through relationships to people and numbers and also the city's geolocation (longitude and latitude), which is important for mapping applications and distance calculations.

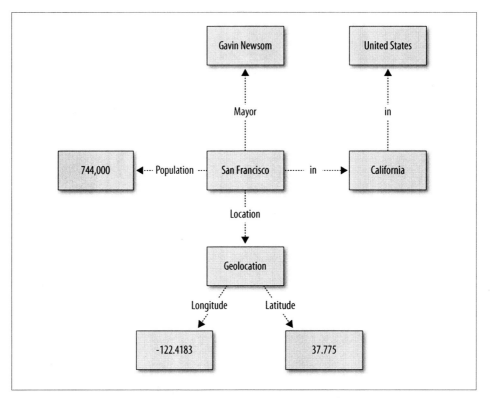

Figure 2-7. An example of location data expressed as a graph

You can download a file containing triples about places from *http://semprog.com/psw/chapter2/place_triples.txt*. In a Python session, load it up and try some simple queries:

```
>>> from simplegraph import SimpleGraph
>>> placegraph=SimpleGraph()
>>> placegraph.loadfile("place_triples.txt")
```

This pattern returns everything we know about San Francisco:

```
>>> for t in placegraph.triples((None,"name","San Francisco")):
...     print t
...
(u'San_Francisco_California', 'name', 'San Francisco')
>>> for t in placegraph.triples(("San_Francisco_California",None,None)):
...     print t
...
('San_Francisco_California', u'name', u'San Francisco')
('San_Francisco_California', u'inside', u'California')
('San_Francisco_California', u'longitude', u'-122.4183')
('San_Francisco_California', u'latitude', u'37.775')
('San_Francisco_California', u'mayor', u'Gavin Newsom')
('San_Francisco_California', u'population', u'744042')
```

This pattern shows all the mayors in the graph:

```
>>> for t in placegraph.triples((None,'mayor',None)):
...     print t
...
(u'Aliso_Viejo_California', 'mayor', u'Donald Garcia')
(u'San_Francisco_California', 'mayor', u'Gavin Newsom')
(u'Hillsdale_Michigan', 'mayor', u'Michael Sessions')
(u'San_Francisco_California', 'mayor', u'John Shelley')
(u'Alameda_California', 'mayor', u'Lena Tam')
(u'Stuttgart_Germany', 'mayor', u'Manfred Rommel')
(u'Athens_Greece', 'mayor', u'Dora Bakoyannis')
(u'Portsmouth_New_Hampshire', 'mayor', u'John Blalock')
(u'Cleveland_Ohio', 'mayor', u'Newton D. Baker')
(u'Anaheim_California', 'mayor', u'Curt Pringle')
(u'San_Jose_California', 'mayor', u'Norman Mineta')
(u'Chicago_Illinois', 'mayor', u'Richard M. Daley')

...
```

We can also try something a little bit more sophisticated by using a loop to get all the cities in California and then getting their mayors:

```
>>> cal_cities=[p[0] for p in placegraph.triples((None,'inside','California'))]
>>> for city in cal_cities:
...     for t in placegraph.triples((city,'mayor',None)):
...         print t
...
(u'Aliso_Viejo_California', 'mayor', u'William Phillips')
(u'Chula_Vista_California', 'mayor', u'Cheryl Cox')
(u'San_Jose_California', 'mayor', u'Norman Mineta')
(u'Fontana_California', 'mayor', u'Mark Nuaimi')
(u'Half_Moon_Bay_California', 'mayor', u'John Muller')
(u'Banning_California', 'mayor', u'Brenda Salas')
(u'Bakersfield_California', 'mayor', u'Harvey Hall')
(u'Adelanto_California', 'mayor', u'Charley B. Glasper')
(u'Fresno_California', 'mayor', u'Alan Autry')
(etc...)
```

This is a simple example of joining data in multiple steps. As mentioned previously, the next chapter will show you how to build a simple graph-query language to do all of this in one step.

Celebrities

Our next example is a fun one: celebrities. The wonderful thing about famous people is that other people are always talking about what they're doing, particularly when what they're doing is unexpected. For example, take a look at the graph around the ever-controversial Britney Spears, shown in Figure 2-8.

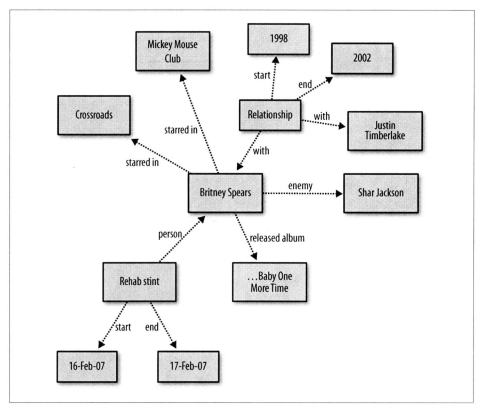

Figure 2-8. An example of celebrity data expressed as a graph

Even from this very small section of Ms. Spears's life, it's clear that there are lots of different things and, more importantly, lots of different *types* of things we say about celebrities. It's almost comical to think that one could frontload the schema design of everything that a famous musician or actress might do in the future that would be of interest to people. This graph has already failed to include such things as favorite nightclubs, estranged children, angry head-shavings, and cosmetic surgery controversies.

We've created a sample file of triples about celebrities at *http://semprog.com/psw/chapter2/celeb_triples.txt*. Feel free to download this, load it into a graph, and try some fun examples:

```
>>> from simplegraph import SimpleGraph
>>> cg=SimpleGraph()
>>> cg.load('celeb_triples.csv')
>>> jt_relations=[t[0] for t in cg.triples((None,'with','Justin Timberlake'))]
>>> jt_relations # Justin Timberlake's relationships
[u'rel373', u'rel372', u'rel371', u'rel323', u'rel16', u'rel15',
 u'rel14', u'rel13', u'rel12', u'rel11']
```

```
>>> for rel in jt_relations:
...     print [t[2] for t in cg.triples((rel,'with',None))]
...
[u'Justin Timberlake', u'Jessica Biel']
[u'Justin Timberlake', u'Jenna Dewan']
[u'Justin Timberlake', u'Alyssa Milano']
[u'Justin Timberlake', u'Cameron Diaz']
[u'Justin Timberlake', u'Britney Spears']
[u'Justin Timberlake', u'Jessica Biel']
[u'Justin Timberlake', u'Jenna Dewan']
[u'Justin Timberlake', u'Alyssa Milano']
[u'Justin Timberlake', u'Cameron Diaz']
[u'Justin Timberlake', u'Britney Spears']
>>> bs_movies=[t[2] for t in cg.triples(('Britney Spears','starred_in',None))]
>>> bs_movies # Britney Spears' movies
[u'Longshot', u'Crossroads', u"Darrin's Dance Grooves", u'Austin Powers: Goldmember']
>> movie_stars=set()
>>> for t in cg.triples((None,'starred_in',None)):
...     movie_stars.add(t[0])
...
>>> movie_stars # Anyone with a 'starred_in' assertion
set([u'Jenna Dewan', u'Cameron Diaz', u'Helena Bonham Carter', u'Stephan Jenkins',
u'Pen\xe9lope Cruz', u'Julie Christie', u'Adam Duritz', u'Keira Knightley',
(etc...)
```

As an exercise, see if you can write Python code to answer some of these questions:

- Which celebrities have dated more than one movie star?

- Which musicians have spent time in rehab? (Use the person predicate from rehab nodes.)

- Think of a new predicate to represent fans. Add a few assertions about stars of whom you are a fan. Now find out who your favorite stars have dated.

Hopefully you're starting to get a sense of not only how easy it is to add new assertion types to a triplestore, but also how easy it is to try things with assertions created by someone else. You can start asking questions about a dataset from a single file of triples. The essence of semantic data is ease of extensibility and ease of sharing.

Business

Lest you think that semantic data modeling is all about movies, entertainment, and celebrity rivalries, Figure 2-9 shows an example with a little more gravitas: data from the business world. This graph shows several different types of relationships, such as company locations, revenue, employees, directors, and political contributions.

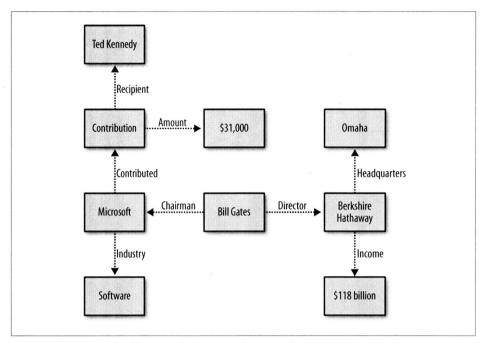

Figure 2-9. An example of business data expressed as a graph

Obviously, a lot of information that is particular to these businesses could be added to this graph. The relationships shown here are actually quite generic and apply to most companies, and since companies can do so many different things, it's easy to imagine more specific relationships that could be represented. We might, for example, want to know what investments Berkshire Hathaway has made, or what software products Microsoft has released. This just serves to highlight the importance of a flexible schema when dealing with complex domains such as business.

Again, we've provided a file of triples for you to download, at *http://semprog.com/psw/ chapter2/business_triples.csv*. This is a big graph, with 36,000 assertions about 3,000 companies.

Here's an example session:

```
>>> from simplegraph import SimpleGraph
>>> bg=SimpleGraph()
>>> bg.load('business_triples.csv')

>>> # Find all the investment banks
>>> ibanks=[t[0] for t in bg.triples((None,'industry','Investment Banking'))]
>>> ibanks
[u'COWN', u'GBL', u'CLMS', u'WDR', u'SCHW', u'LM', u'TWPG', u'PNSN', u'BSC', u'GS',
 u'NITE', u'DHIL', u'JEF', u'BLK', u'TRAD', u'LEH', u'ITG', u'MKTX', u'LAB', u'MS',
 u'MER', u'OXPS', u'SF']
```

```
>>> bank_contrib={}  # Contribution nodes from Investment banks
>>> for b in ibanks:
...     bank_contrib[b]=[t[0] for t in bg.triples((None,'contributor',b))]

>>> # Contributions from investment banks to politicians
>>> for b,contribs in bank_contrib.items():
...     for contrib in contribs:
...             print [t[2] for t in bg.triples((contrib,None,None))]

[u'BSC', u'30700.0', u'Orrin Hatch']
[u'BSC', u'168335.0', u'Hillary Rodham Clinton']
[u'BSC', u'5600.0', u'Christopher Shays']
[u'BSC', u'5000.0', u'Barney Frank']
(etc...)

>>> sw=[t[0] for t in bg.triples((None,'industry','Computer software'))]
>>> sw
[u'GOOG', u'ADBE', u'IBM',   # Google, Adobe, IBM

>>> # Count locations
>>> locations={}
>>> for company in sw:
...     for t in bg.triples((company,'headquarters',None)):
...             locations[t[2]]=locations.setdefault(t[2],0)+1

>>> # Locations with 3 or more software companies
>>> [loc for loc,c in locations.items() if c>=3]
[u'Austin_Texas', u'San_Jose_California', u'Cupertino_California',
u'Seattle_Washington']
```

Notice that we've used ticker symbols as IDs, since they are guaranteed to be unique and short.

We hope this gives you a good idea of the many different types of data that can be expressed in triples, the different things you might ask of the data, and how easy it is to extend it. Now let's move on to creating a better method of querying.

## CHAPTER 3

Using Semantic Data

So far, you've seen how using explicit semantics can make it easier to share your data and extend your existing system as you get new data. In this chapter we'll show that semantics also makes it easier to develop reusable techniques for querying, exploring, and using data. Capturing semantics in the data itself means that reconfiguring an algorithm to work on a new dataset is often just a matter of changing a few keywords.

We'll extend the simple triplestore we built in Chapter 2 to support constraint-based querying, simple feed-forward reasoning, and graph searching. In addition, we'll look at integrating two graphs with different kinds of data but create separate visualizations of the data using tools designed to work with semantic data.

A Simple Query Language

Up to this point, our query methods have looked for patterns within a single triple by setting the subject, predicate, or object to a wildcard. This is useful, but by treating each triple independently, we aren't able to easily query across a graph of relationships. It is these graph relationships, spanning multiple triples, that we are most interested in working with.

For instance, in Chapter 2 when we wanted to discover which mayors served cities in California, we were forced to run one query to find "cities" (subject) that were "inside" (predicate) "California" (object) and then independently loop through all the cities returned, searching for triples that matched the "city" (subject) and "mayor" (predicate).

To simplify making queries like this, we will abstract this basic query process and develop a simple language for expressing these types of graph relationships. This graph pattern language will form the basis for building more sophisticated queries and applications in later chapters.

Variable Binding

Let's consider a fragment of the graph represented by the `places_triples` used in the section "Other Examples" on page 29. We can visualize this graph, representing three cities and the predicates "inside" and "mayor", as shown in Figure 3-1.

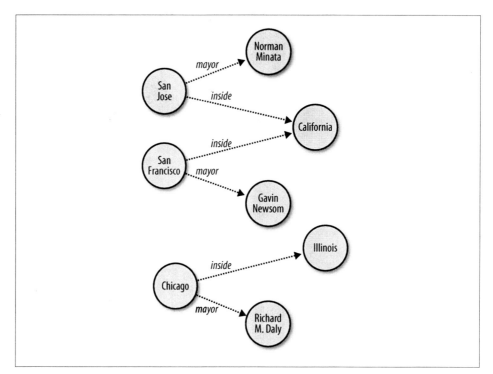

Figure 3-1. A graph of city mayors and locations

Consider the two statements about San Francisco: it is "inside" California, and Gavin Newsom is the mayor. When we express this piece of the graph in triples, we use a common subject identifier `San_Francisco_California` to indicate that the statements describe a single entity:

```
("San_Francisco_California", "inside", "California")
("San_Francisco_California", "mayor", "Gavin Newsom")
```

The shared subject identifier `San_Francisco_California` indicates that the two statements are about the same entity, the city of San Francisco. When an identifier is used multiple times in a set of triples, it indicates that a node in the graph is shared by all the assertions. And just as you are free to select any name for a variable in a program, the choice of identifiers used in triples is arbitrary, too. As long as you are consistent in your use of an identifier, the facts that you can learn from the assertions will be clear.

For instance, we could have said Gavin Newsom is the mayor of a location within California with the triples:

```
("foo", "inside", "California")
("foo", "mayor", "Gavin Newsom")
```

While San_Francisco_California is a useful moniker to help humans understand that "San Francisco" is the location within California that has Newsom as a mayor, the relationships described in the triples using San_Francisco_California and foo are identical.

 It is important not to think of the identifier as the "name" of an entity. Identifiers are simply arbitrary symbols that indicate multiple assertions are related. As in the original places_triples dataset in Chapter 2, if we wanted to name the entity where Newsom is mayor, we would need to make the name relationship explicit with another triple:

```
("foo", "name", "San Francisco")
```

With an understanding of how shared identifiers are used to indicate shared nodes in the graph, we can return to our question of which mayors serve in California. As before, we start by asking, "Which nodes participate in an assertion with the predicate of 'inside' and object of 'California'?" Using our current query method, we would write this constraint as this triple query:

```
(None, "inside", "California")
```

and our query method would return the set of triples that matched the pattern.

Instead of using a wildcard, let's introduce a variable named ?city to collect the identifiers for the nodes in the graph that satisfy our constraints. For the graph fragment pictured in Figure 3-1 and the triple query pattern ("?city", "inside", "California"), there are two triples that satisfy the constraints, which gives us two possible values for the variable ?city: the identifiers San_Francisco_California and San_Jose_California.

We can express these results as a list of dictionaries mapping the variable name to each matching identifier. We refer to these various possible mappings as "bindings" of the variable ?city to different values. In our example, the results would be expressed as:

```
[{"?city": "San_Francisco_California"}, {"?city": "San_Jose_California"}]
```

We now have a way to assign results from a triple query to a variable. We can use this to combine multiple triple queries and take the intersection of the individual result sets. For instance, this query specifies the intersection of two triple queries in order to find all cities in California that have a mayor named Norman Mineta:

```
("?city", "inside", "California")
("?city", "mayor", "Norman Mineta")
```

The result [{"?city": "San_Jose_California"}] is the solution to this graph query because it is the only value for the variable ?city that is in all of the individual triple query results. We call the individual triple queries in the graph query "constraints" because each triple query constrains and limits the possible results for the whole graph query.

We can also use variables to ask for more results. By adding an additional variable, we can construct a graph query that tells us which mayors serve cities within California:

```
("?city", "inside", "California")
("?city", "mayor", "?name_of_mayor")
```

There are two solutions to this query. In the first solution, the variable ?city is bound to the identifier San_Jose_California, which causes the variable ?name_of_mayor to be bound to the identifier "Norman Mineta". In the second solution, ?city will be bound to San_Francisco_California, resulting in the variable ?name_of_mayor being bound to "Gavin Newsom". We can write this solution set as:

```
[{"?city": "San_Francisco_California", "?name_of_mayor": "Gavin Newsom"},
 {"?city": "San_Jose_California", "?name_of_mayor": "Norman Mineta"}]
```

It is important to note that all variables in a solution are bound simultaneously—each dictionary represents an alternative and valid set of bindings given the constraints in the graph query.

Implementing a Query Language

In this section, we'll show you how to add variable binding to your existing triplestore. This will allow you to ask the complex questions in the previous chapter with a single method call, instead of doing separate loops and lookups for each step. For example, the following query returns all of the investment banks in New York that have given money to Utah Senator Orrin Hatch:

```
>>> bg.query([('?company','headquarters','New_York_NY'),
              ('?company','industry','Investment Banking'),
              ('?company','contributor','?contribution'),
              ('?contribution','recipient','Orrin Hatch'),
              ('?contribution','amount','?dollars')])
```

The variables are preceded with a question mark, and everything else is considered a constant. This call to the query method tries to find possible values for company, contribution, and dollars that fulfill the following criteria:

1. company is headquartered in New York

2. company is in the industry Investment Banking

3. company made a contribution called contribution

4. contribution had a recipient called Orrin Hatch

5. contribution had an amount equal to dollars

From the session at the end of Chapter 2, we know that one possible answer is:

```
{'?company':'BSC',
 '?contribution':'contXXX',
 '?dollars':'30700'}
```

If BSC has made multiple contributions to Orrin Hatch, we'll get a separate solution for each contribution.

Before we get to the implementation, just to make sure you're clear, here's another example:

```
>>> cg.query([('?rel1','with','?person'),
              ('?rel1','with','Britney Spears'),
              ('?rel1','end','?year1'),
              ('?rel2','with','?person'),
              ('?rel2','start','?year1')])
```

This asks, "Which person started a new relationship in the same year that their relationship with Britney Spears ended?" The question is a little convoluted, but hopefully it shows the power of querying with variable binding. In this case, we're looking for sets that fulfill the following criteria:

1. rel1 (a relationship) involved person
2. rel1 also involved Britney Spears
3. rel1 ended in year1
4. rel2 involved person
5. rel2 started in year1

Since there's nothing saying that person can't be Britney Spears, it's possible that she could be one of the answers, if she started a new relationship the same year she ended one. As we look at more sophisticated querying languages, we'll see ways to impose negative constraints.

The implementation of query is a very simple method for variable binding. It's not super efficient and doesn't do any query optimization, but it will work well on the sets we've been working with and should help you understand how variable binding works. You can add the following code to your existing simplegraph class, or you can download *http://semprog.com/psw/chapter3/simplegraph.py*:

```
def query(self,clauses):
    bindings = None
    for clause in clauses:
        bpos = {}
        qc = []
        for pos, x in enumerate(clause):
            if x.startswith('?'):
                qc.append(None)
                bpos[x] = pos
            else:
                qc.append(x)
        rows = list(self.triples((qc[0], qc[1], qc[2])))
```

```
        if bindings == None:
            # This is the first pass, everything matches
            bindings = []
            for row in rows:
                binding = {}
                for var, pos in bpos.items():
                    binding[var] = row[pos]
                bindings.append(binding)
    else:
            # In subsequent passes, eliminate bindings that don't work
            newb = []
            for binding in bindings:
                for row in rows:
                    validmatch = True
                    tempbinding = binding.copy()
                    for var, pos in bpos.items():
                        if var in tempbinding:
                            if tempbinding[var] != row[pos]:
                                validmatch = False
                        else:
                            tempbinding[var] = row[pos]
                    if validmatch: newb.append(tempbinding)
            bindings = newb
    return bindings
```

This method loops over each clause, keeping track of the positions of the variables (any string that starts with ?). It replaces all the variables with None so that it can use the **triples** method already defined in **simplegraph**. It then gets all the rows matching the pattern with the variables removed.

For every row in the set, it looks at the positions of the variables in the clause and tries to fit the values to one of the existing bindings. The first time through there are no existing bindings, so every row that comes back becomes a potential binding. After the first time, each row is compared to the existing bindings—if it matches, more variables are added and the binding is added to the current set. If there are no rows that match an existing binding, that binding is removed.

Try using this new query method in a Python session for the two aforementioned queries:

```
>>> from simplegraph import SimpleGraph()
>>> bg = SimpleGraph()
>>> bg.load('business_triples.csv')
>>> bg.query([('?company','headquarters','New_York_New_York'),
              ('?company','industry','Investment Banking'),
              ('?cont','contributor','?company'),
              ('?cont','recipient','Orrin Hatch'),
              ('?cont','amount','?dollars')])
[{'company': u'BSC', 'cont': u'contrib285', 'dollars': u'30700.0'}]
>>> cg = SimpleGraph()
>>> cg.load('celeb_triples.csv')
>>> cg.query([('?rel1','with','?person'),
              ('?rel1','with','Britney Spears'),
              ('?rel1','end','?year1'),
```

```
            ('?rel2','with','?person'),
            ('?rel2','start','?year1')])
[{'person': u'Justin Timberlake', 'rel1': u'rel16', 'year1': u'2002',
    'rel2': u'rel372'} ...
```

Now see if you can formulate some queries of your own on the movie graph. For example, which actors have starred in movies directed by Ridley Scott as well as movies directed by George Lucas?

Feed-Forward Inference

Inference is the process of deriving new information from information you already have. This can be used in a number of different ways. Here are a few examples of types of inference:

Simple and deterministic
> If I know a rock weighs 1 kg, I can infer that the same rock weighs 2.2 lbs.

Rule-based
> If I know a person is under 16 and in California, I can infer that they are not allowed to drive.

Classifications
> If I know a company is in San Francisco or Seattle, I can classify it as a "west coast company."

Judgments
> If I know a person's height is 6 feet or more, I refer to them as tall.

Online services
> If I know a restaurant's address, I can use a geocoder to find its coordinates on a map.

Obviously, the definition of what counts as "information" and which rules are appropriate will vary depending on the context, but the idea is that by using rules along with some knowledge and outside services, we can generate new assertions from our existing set of assertions.

The last example is particularly interesting in the context of web development. The Web has given us countless online services that can be queried programmatically. This means it is possible to take the assertions in a triplestore, formulate a request to a web service, and use the results from the query to create new assertions. In this section, we'll show examples of basic rule-based inference and of using online services to infer new triples from existing ones.

Inferring New Triples

The basic pattern of inference is simply to query for information (in the form of bindings) relevant to a rule, then apply a transformation that turns these bindings into a new set of triples that get added back to the triplestore. We're going to create a basic class that defines inference rules, but first we'll add a new method for applying rules to the SimpleGraph class. If you downloaded *http://semprog.com/psw/chapter3/simple graph.py*, you should already have this method. If not, add it to your class:

```
def applyinference(self,rule):
    queries = rule.getqueries()
    bindings=[]
    for query in queries:
        bindings += self.query(query)
    for b in bindings:
        new_triples = rule.maketriples(b)
        for triple in new_triples:
            self.add(triple)
```

This method takes a rule (usually an instance of InferenceRule) and runs its query to get a set of bindings. It then calls rule.maketriples on each set of bindings, and adds the returned triples to the store.

The InferenceRule class itself doesn't do much, but it serves as a base from which other rules can inherit. Child classes will override the getquery method and define a new method called _maketriples, which will take each binding as a parameter. You can download *http://semprog.com/psw/chapter3/inferencerule.py*, which contains the class definition of InferenceRule and the rest of the code from this section. If you prefer, you can create a new file called *inferencerule.py* and add the following code:

```
class InferenceRule:
    def getqueries(self):
        return []

    def maketriples(self,binding):
        return self._maketriples(**binding)
```

Now we are ready to define a new rule. Our first rule will identify companies headquartered in cities on the west coast of the United States and identify them as such:

```
class WestCoastRule(InferenceRule):
    def getqueries(self):
        sfoquery = [('?company', 'headquarters', 'San_Francisco_California')]
        seaquery = [('?company', 'headquarters', 'Seattle_Washington')]
        laxquery = [('?company', 'headquarters', 'Los_Angelese_California')]
        porquery = [('?company', 'headquarters', 'Portland_Oregon')]
        return [sfoquery, seaquery, laxquery, porquery]

    def _maketriples(self, company):
        return [(company, 'on_coast', 'west_coast')]
```

The rule class lets you define several queries: in this case, there's a separate query for each of the major west coast cities. The variables used in the queries themselves (in this case, just company) become the parameters for the call to _maketriples. This rule asserts that all the companies found by the first set of queries are on the west coast.

You can try this in a session:

```
>>> wcr = WestCoastRule()
>>> bg.applyinference(wcr)
>>> list(bg.triples((None, 'on_coast', None)))
[(u'PCL', 'on_coast', 'west_coast'), (u'PCP', 'on_coast', 'west_coast') ...
```

Although we have four different queries for our west coast rule, each of them has only one clause, which makes the rules themselves very simple. Here's a rule with a more sophisticated query that can be applied to the celebrities graph:

```
class EnemyRule(InferenceRule):
    def getqueries(self):
        partner_enemy = [('?person', 'enemy', '?enemy'),
                         ('?rel', 'with', '?person'),
                         ('?rel', 'with', '?partner')]
        return [partner_enemy]

    def _maketriples(self, person, enemy, rel, partner):
        return (partner, 'enemy', enemy)
```

Just from looking at the code, can you figure out what this rule does? It asserts that if a **person** has a relationship **partner** and also an **enemy**, then their partner has the same enemy. That may be a little presumptuous, but nonetheless it demonstrates a simple logical rule that you can make with this pattern. Again, try it in a session:

```
>>> from simplegraphq import *
>>> cg = SimpleGraph()
>>> cg.load('celeb_triples.csv')
>>> er = EnemyRule()
>>> list(cg.triples((None, 'enemy', None)))
[(u'Jennifer Aniston', 'enemy', u'Angelina Jolie')...
>>> cg.applyinference(er)
>>> list(cg.triples((None, 'enemy', None)))
[(u'Jennifer Aniston', 'enemy', u'Angelina Jolie'), (u'Vince Vaughn', 'enemy', \
    u'Angelina Jolie')...
```

These queries are all a little self-directed, since they operate directly on the data and have limited utility. We'll now see how we can use this reasoning framework to add more information to the triplestore by retrieving it from outside sources.

Geocoding

Geocoding is the process of taking an address and getting the geocoordinates (the longitude and latitude) of that address. When you use a service that shows the location of an address on a map, the address is always geocoded first. Besides displaying on a map, geocoding is useful for figuring out the distance between two places and determining automatically if an address is inside or outside particular boundaries. In this section, we'll show you how to use an external geocoder to infer triples from addresses of businesses.

Using a free online geocoder

There are a number of geocoders that are accessible through an API, many of which are free for limited use. We're going to use the site *http://geocoder.us/*, which provides free geocoding for noncommercial use and very cheap access for commercial use for addresses within the United States. The site is based on a Perl script called *Geo::Coder::US*, which can be downloaded from *http://search.cpan.org/~sderle/Geo -Coder-US/* if you're interested in hosting your own geocoding server.

There are several other commercial APIs, such as Yahoo!'s, that offer free geocoding with various limitations and can be used to geocode addresses in many different countries. We've provided alternative implementations of the geocoding rule on our site *http://semprog.com/*.

The free service of geocoder.us is accessed using a REST API through a URL that looks like this:

http://rpc.geocoder.us/service/csv?address=1600+Pennsylvania+Avenue,+Washing ton+DC

If you go to this URL in your browser, you should get a single line response:

```
38.898748,-77.037684,1600 Pennsylvania Ave NW,Washington,DC,20502
```

This is a comma-delimited list of values showing the latitude, longitude, street address, city, state, and ZIP code. Notice how, in addition to providing coordinates, the geocoder has also changed the address to a more official "Pennsylvania Ave NW" so that you can easily compare addresses that were written in different ways. Try a few other addresses and see how the results change.

Adding a geocoding rule

To create a geocoding rule, just make a simple class that extends `InferenceRule`. The query for this rule finds all triples that have "address" as their predicate; then for each address it contacts the geocoder and tries to extract the latitude and longitude:

```python
from urllib import urlopen, quote_plus

class GeocodeRule(InferenceRule):
    def getquery(self):
        return [('?place', 'address', '?address')]

    def _maketriples(self, place, address):
        url = 'http://rpc.geocoder.us/service/csv?address=%s' % quote_plus(address)
        con = urlopen(url)
        data = con.read()
        con.close()
        parts = data.split(',')
        if len(parts) >= 5:
            return [(place, 'longitude', parts[0]),
                    (place, 'latitude', parts[1])]
        else:
```

```
        # Couldn't geocode this address
        return []
```

You can then run the geocoder by creating a new graph and putting an address in it:

```
>>> from simplegraph import *
>>> from inferencerule import *
>>> geograph = SimpleGraph()
>>> georule = GeocodeRule()
>>> geograph.add(('White House', 'address', '1600 Pennsylvania Ave, Washington, DC'))
>>> list(geograph.triples((None, None, None))
[('White House', 'address', '1600 Pennsylvania Ave, Washington, DC')]
>>> geograph.applyinference(georule)
>>> list(geograph.triples((None, None, None))
[('White House', 'latitude', '-77.037684'),
 ('White House', 'longitude', '38.898748'),
 ('White House', 'address', '1600 Pennsylvania Ave, Washington, DC')]
```

We've provided a file containing a couple of restaurants in Washington, DC, at *http://semprog.com/psw/chapter3/DC_addresses.csv*. Try downloading the data and running it through the geocoding rule. This may take a minute or two, as access to the geocoder is sometimes throttled. If you're worried it's taking too long, you can add print statements to the geocode rule to monitor the progress.

Chains of Rules

The fact that these inferences create new assertions in the triplestore is incredibly useful to us. It means that we can write inference rules that operate on the results of other rules without explicitly coordinating all of the rules together. This allows the creation of completely decoupled, modular systems that are very robust to change and failure.

To understand how this works, take a look at CloseToRule. This takes a place name and a graph in its constructor. It queries for every geocoded item in the graph, calculates how far away they are, and then asserts close_to for all of those places that are close by:

```
class CloseToRule(InferenceRule):

    def __init__(self, place, graph):
        self.place = place
        laq = list(graph.triples((place, 'latitude', None)))
        loq = list(graph.triples((place, 'longitude', None)))

        if len(laq) == 0 or len(loq) == 0:
            raise "Exception","%s is not geocoded in the graph" % place

        self.lat = float(laq[0][2])
        self.long = float(loq[0][2])

    def getqueries(self):
        geoq=[('?place', 'latitude', '?lat'), ('?place', 'longitude', '?long')]
        return [geoq]

    def _maketriples(self, place, lat, long):
```

```
# Formula for distance in miles from geocoordinates
distance=((69.1*(self.lat - float(lat)))**2 + \
    (53*(self.lat - float(lat)))**2)**.5

# Are they less than a mile apart
if distance < 1:
    return [(self.place, 'close_to', place)]
else:
    return [(self.place, 'far_from', place)]
```

Now you can use these two rules together to first geocode the addresses, then create assertions about which places are close to the White House:

```
>>> from simplegraph import *
>>> from inferencerule import *
>>> pg = SimpleGraph()
>>> pg.load('DC_addresses.csv')
>>> georule = GeocodeRule()
>>> pg.applyinference(georule)
>>> whrule = CloseToRule('White House',pg)
>>> pg.applyinference(whrule)
>>> list(pg.triples((None, 'close_to', None)))
[('White House', 'close_to', u'White House'),
 ('White House', 'close_to', u'Pot Belly'),
 ('White House', 'close_to', u'Equinox')]
```

Now we've chained together two rules. Keep this session open while we take a look at another rule, which identifies restaurants that are likely to be touristy. In the DC_addresses file, there are assertions about what kind of place each venue is ("tourist attraction" or "restaurant") and also information about prices ("cheap" or "expensive"). The TouristyRule combines all this information along with the close_to assertions to determine which restaurants are touristy:

```
class TouristyRule(InferenceRule):
    def getqueries(self):
        tr = [('?ta', 'is_a', 'Tourist Attraction'),
              ('?ta', 'close_to', '?restaurant'),
              ('?restaurant', 'is_a', 'restaurant'),
              ('?restaurant', 'cost', 'cheap')]

    def _maketriples(self, ta, restaurant):
        return [(restaurant, 'is_a', 'touristy restaurant')]
```

That is, if a restaurant is cheap and close to a tourist attraction, then it's probably a pretty touristy restaurant:

```
>>> tr = TouristyRule()
>>> pg.applyinference(tr)
>>> list(pg.triples((None, 'is_a', 'touristy restaurant')))
[(u'Pot Belly', 'is_a', 'touristy restaurant')]
```

So we've gone from a bunch of addresses and restaurant prices to predictions about restaurants where you might find a lot of tourists. You can think of this as a group of dependent functions, similar to Figure 3-2, which represents the way we normally think about data processing.

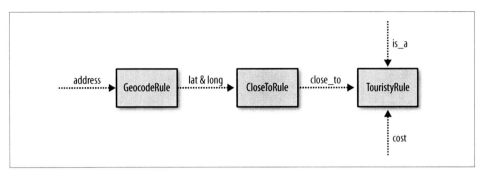

Figure 3-2. A "chain view" of the rules

What's important to realize here is that the rules exist totally independently. Although we ran the three rules in sequence, they weren't aware of each other—they just looked to see if there were any triples that they knew how to deal with and then created new ones based on those. These rules can be run continuously—even from different machines that have access to the same triplestore—and still work properly, and new rules can be added at any time. To help understand what this implies, consider a few examples:

1. Geographical oddities may have a latitude and longitude but no address. They can be put right into the triplestore, and CloseToRule will find them without them ever being noticed by GeocodeRule.

2. We may invent new rules that add addresses to the database, which will run through the whole chain.

3. We may initially know about a restaurant's existence but not know its cost. In this case, GeocodeRule can geocode it, CloseToRule can assert that it is close to things, but TouristyRule won't be able to do anything with it. However, if we later learn the cost, TouristyRule can be activated *without activating the other rules*.

4. We may know that some place is close to another place but not know its exact location. Perhaps someone told us that they walked from the White House to a restaurant. This information can be used by TouristyRule without requiring the other rules to be activated.

So a better way to think about this is Figure 3-3.

The rules all share an information store and look for information they can use to generate new information. This is sometimes referred to as a *multi-agent blackboard*. It's a different way of thinking about programming that sacrifices in efficiency but that has the advantages of being very decoupled, easily distributable, fault tolerant, and flexible.

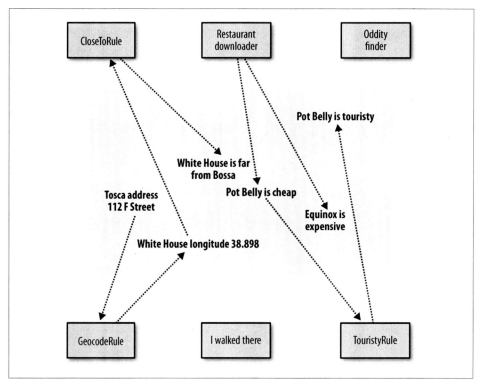

Figure 3-3. Inference rules reading from and writing to a "blackboard"

A Word About "Artificial Intelligence"

It's important to realize, of course, that "intelligence" doesn't come from chains of symbolic logic like this. In the past, many people have made the mistake of trying to generate intelligent behavior this way, attempting to model large portions of human knowledge and reasoning processes using only symbolic logic. These approaches always reveal that there are pretty severe limitations on what can be modeled. Predicates are imprecise, so many inferences are impossible or are only correct in the context of a particular application.

The examples here show triggers for making new assertions entirely from existing ones, and also show ways to query other sources to create new assertions based on existing ones.

Searching for Connections

A common question when working with a graph of data is how two entities are connected. The most common algorithm for finding the shortest path between two points

---

in a graph is an algorithm called *breadth-first search*. The breadth-first search algorithm finds the shortest path between two nodes in a graph by taking the first node, looking at all of its neighbors, looking at all of the neighbors of its neighbors, and so on until the second node is found or until there are no new nodes to look at. The algorithm is guaranteed to find the shortest path if one exists, but it may have to look at all of the edges in the graph in order to find it.

Six Degrees of Kevin Bacon

A good example of the breadth-first search algorithm is the trivia game "Six Degrees of Kevin Bacon," where one player names a film actor, and the other player tries to connect that actor to the actor Kevin Bacon through the shortest path of movies and co-stars. For instance, one answer for the actor Val Kilmer would be that Val Kilmer starred in *Top Gun* with Tom Cruise, and Tom Cruise starred in *A Few Good Men* with Kevin Bacon, giving a path of length 2 because two movies had to be traversed. See Figure 3-4.

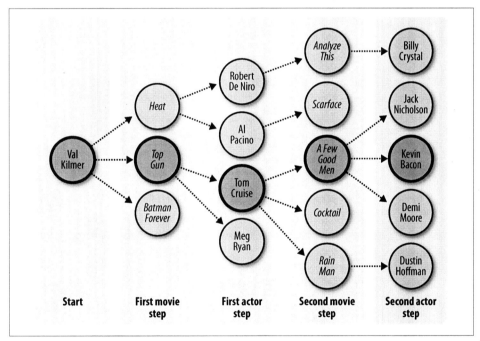

Figure 3-4. Breadth-first search from Val Kilmer to Kevin Bacon

Here's an implementation of breadth-first search over the movie data introduced in Chapter 2. On each iteration of the `while` loop, the algorithm processes a successive group of nodes one edge further than the last. So on the first iteration, the starting actor node is examined, and all of its adjacent movies that have not yet been seen are added to the `movieIds` list. Then all of the actors in those movies are found and are added to the `actorIds` list if they haven't yet been seen. If one of the actors is the one we are looking for, the algorithm finishes.

The shortest path back to the starting node from each examined node is stored at each step as well. This is done in the "parent" variable, which at each step points to the node that was used to find the current node. This path of "parent" nodes can be followed back to the starting node:

```
def moviebfs(startId, endId, graph):
    actorIds = [(startId, None)]
    # keep track of actors and movies that we've seen:
    foundIds = set()
    iterations = 0
    while len(actorIds) > 0:
        iterations += 1
        print "Iteration " + str(iterations)
        # get all adjacent movies:
        movieIds = []
        for actorId, parent in actorIds:
            for movieId, _, _ in graph.triples((None, "starring", actorId)):
                if movieId not in foundIds:
                    foundIds.add(movieId)
                    movieIds.append((movieId, (actorId, parent)))
        # get all adjacent actors:
        nextActorIds = []
        for movieId, parent in movieIds:
            for _, _, actorId in graph.triples((movieId, "starring", None)):
                if actorId not in foundIds:
                    foundIds.add(actorId)
                    # we found what we're looking for:
                    if actorId == endId: return (iterations, (actorId, \
                        (movieId, parent)))
                    else: nextActorIds.append((actorId, (movieId, parent)))
        actorIds = nextActorIds
    # we've run out of actors to follow, so there is no path:
    return (None, None)
```

Now we can define a function that runs `moviebfs` and recovers the shortest path:

```
def findpath(start, end, graph):
    # find the ids for the actors and compute bfs:
    startId = graph.value(None, "name", start)
    endId = graph.value(None, "name", end)
    distance, path = moviebfs(startId, endId, graph)
    print "Distance: " + str(distance)
    # walk the parent path back to the starting node:
    names = []
    while path is not None:
        id, nextpath = path
        names.append(graph.value(id, "name", None))
        path = nextpath
    print "Path: " + ", ".join(names)
```

Here's the output for Val Kilmer, Bruce Lee, and Harrison Ford. (Note that our movie data isn't as complete as some databases on the Internet, so there may in fact be a shorter path for some of these actors.)

```
>>> import simplegraph
>>> graph = simplegraph.SimpleGraph()
>>> graph.load("movies.csv")
>>> findpath("Val Kilmer", "Kevin Bacon", graph)
Iteration 1
Iteration 2
Distance: 2
Path: Kevin Bacon, A Few Good Men, Tom Cruise, Top Gun, Val Kilmer
>>> findpath("Bruce Lee", "Kevin Bacon", graph)
Iteration 1
Iteration 2
Iteration 3
Distance: 3
Path: Kevin Bacon, The Woodsman, David Alan Grier, A Soldier's Story, Adolph Caesar, \
    Fist of Fear, Touch of Death, Bruce Lee
>>> findpath("Harrison Ford", "Kevin Bacon", graph)
Iteration 1
Iteration 2
Distance: 2
Path: Kevin Bacon, Apollo 13, Kathleen Quinlan, American Graffiti, Harrison Ford
```

Shared Keys and Overlapping Graphs

We've talked a lot about the importance of semantics for data integration, but so far we've only shown how you can create and extend separate data silos. But expressing your data this way really shines when you're able to take graphs from two different places and merge them together. Finding a set of nodes where the graphs overlap and linking them together by combining those nodes greatly increases the potential expressiveness of your queries.

Unfortunately, it's not always easy to join graphs, since figuring out which nodes are the same between the graphs is not a trivial matter. There can be misspellings of names, different names for the same thing, or the same name for different things (this is especially true when looking at datasets of people). In later chapters we'll explore the merging problem in more depth. There are certain things that make easy connection points, because they are specific and unambiguous. Locations are a pretty good example, because when I say "San Francisco, California", it almost certainly means the same place to everyone, particularly if they know that I'm currently in the United States.

Example: Joining the Business and Places Graphs

The triples provided for the business and places graphs both contain city names referring to places in the United States. To make the join easy, we already normalized the city names so that they're all written as *City_State*, e.g., San_Francisco_California or

`Omaha_Nebraska`. In a session, you can look at assertions that are in the two separate graphs:

```
>>> from simplegraph import SimpleGraph
>>> bg=SimpleGraph()
>>> bg.load('business_triples.csv')
>>> pg=SimpleGraph()
>>> pg.load('place_triples.csv')

>>> for t in bg.triples((None, 'headquarters', 'San_Francisco_California')):
...     print t
(u'URS', 'headquarters', 'San_Francisco_California')
(u'PCG', 'headquarters', 'San_Francisco_California')
(u'CRM', 'headquarters', 'San_Francisco_California')
(u'CNET', 'headquarters', 'San_Francisco_California')
(etc...)

>>> for t in pg.triples(('San_Francisco_California', None, None)):
...     print t
('San_Francisco_California', u'name', u'San Francisco')
('San_Francisco_California', u'inside', u'California')
('San_Francisco_California', u'longitude', u'-122.4183')
('San_Francisco_California', u'latitude', u'37.775')
('San_Francisco_California', u'mayor', u'Gavin Newsom')
('San_Francisco_California', u'population', u'744042')
```

We're going to merge the data from the places graph, such as population and location, into the business graph. This is pretty straightforward—all we need to do is generate a list of places where companies are headquartered, find the equivalent nodes in the places graph, and copy over all their triples. Try this in your Python session:

```
>>> hq=set([t[2] for t in bg.triples((None,'headquarters',None))])
>>> len(hq)
889
>>> for pt in pg.triples((None, None, None)):
...     if pt[0] in hq: bg.add(pt)
```

Congratulations—you've successfully integrated two distinct datasets! And you didn't even have to worry about their schemas.

Querying the Joined Graph

This may not seem like much yet, but you can now construct queries across both graphs, using the constraint-based querying code from this chapter. For example, you can now get a summary of places where software companies are headquartered:

```
>>> results = bg.query([('?company', 'headquarters', '?city'),
                        ('?city', 'inside', '?region'),
                        ('?company', 'industry', 'Computer software')])
>>> [r['region'] for r in results]
[u'San_Francisco_Bay_Area', u'California', u'United_States', u'Silicon_Valley', \
    u'Northern_California' ...
```

You can also search for investment banks that are located in cities with more than 1,000,000 people:

```
>>> results = bg.query([('?company', 'headquarters', '?city'),
                         ('?city', 'population', '?pop'),
                         ('?company', 'industry', 'Investment banking')])
>>> [r for r in results if int(r['pop']) > 1000000]
[{'city': u'Chicago_Illinois', 'company': u'CME', 'pop': u'2833321'}, ...
```

This is just a small taste of what's possible. Later we'll see how the "semantic web" envisions merging thousands of graphs, which will allow for extremely sophisticated queries across data from many sources.

Basic Graph Visualization

Although we've described the data as conceptually being a graph, and we've been using graphs for explanations, so far we've only looked at the "graphs" as lists of triples. Viewing the contents of the triplestore visually can make it easier to understand and interpret your data. In this section, we'll show you how to use the free software Graphviz to convert the semantic data in your triplestore to images with nodes and edges. We'll show how to view the actual triples and also how to make the graph more concise by graphing the results of a query.

Graphviz

Graphviz is a free software package created by AT&T. It takes simple files describing nodes and connections and applies layout and edge-routing algorithms to produce an image in one of several possible formats. You can download Graphviz from *http://www .graphviz.org/*.

The input files are in a format called DOT, which is a simple text format that looks like this:

```
graph "test" {
    A -- B;
    A -- C;
    C -- D;
}
```

This graph has four nodes: A, B, C, and D. A is connected to B and C, and C is connected to D. There are two different layout programs that come with Graphviz, called dot and neato. Dot produces hierarchical layouts for trees, and neato (which we'll be using in this section) produces layouts for nonhierarchical graphs. The output from neato for the previous example is shown in Figure 3-5.

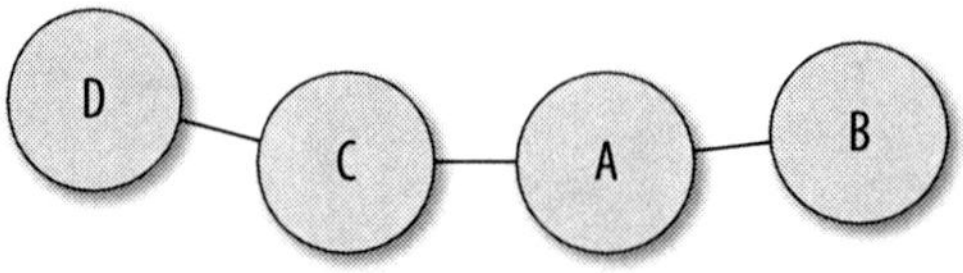

Figure 3-5. Visualization of the ABCD graph

There are a lot of different settings for DOT files, to adjust things like node size, edge thickness, colors, and layout options. We won't cover them all here, but you can find a complete description of the file format in the documentation at *http://www.graphviz .org/pdf/dotguide.pdf*.

Displaying Sets of Triples

The first thing we want to try is creating a graph of a subset of triples. You can download the code for this section at *http://semprog.com/psw/chapter3/graphtools.py*; alternatively, you can create a file called *graphtools.py* and add the `triplestodot` function:

```
def triplestodot(triples, filename):
    out=file(filename, 'w')
    out.write('graph "SimpleGraph" {\n')

    out.write('overlap = "scale";\n')
    for t in triples:
        out.write('"%s" -- "%s" [label="%s"]\n' % (t[0].encode('utf-8'),
                                                   t[2].encode('utf-8'),
                                                   t[1].encode('utf-8')))
    out.write('}')
```

This function simply takes a set of triples, which you would retrieve using the `triples` method of any graph. It creates a DOT file containing those same triples, to be used by Graphviz. Let's try it on the celebrity graph using triples about relationships. In your Python session:

```
>>> from simplegraph import *
>>> from graphtools import *
>>> cg = SimpleGraph()
>>> cg.load('celeb_triples.csv')
>>> rel_triples = cg.triples((None, 'with', None))
>>> triplestodot(rel_triples, 'dating_triples.dot')
```

This should save a file called *dating_triples.dot*. To convert it to an image, you need to run neato from the command line:

```
$ neato -Teps -Odating_triples dating_triples.dot
```

This will save an Encapsulated PostScript (EPS) file called *dating_triples.eps*, which will look something like what you see in Figure 3-6.

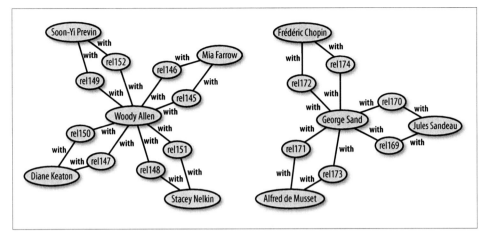

Figure 3-6. Visualization of raw triples in the celebrity dating set

Try generating other images, perhaps a completely unfiltered version that contains every assertion in the graph. You can also try generating images of the other sample data that we've provided.

Displaying Query Results

The problem with graphing the dating triples is that although the graph shows the exact structure of the data, the "rel" nodes shown don't offer any additional information and simply clutter the graph. If we're willing to assume that a relationship is always between two people, then we can eliminate those nodes entirely and connect the people directly to one another. This is pretty easy to do using the query language that we devised at the start of this chapter. The file *graphtools.py* also contains a method called querytodot, which takes a query and two variable names:

```
def querytodot(graph, query, b1, b2, filename):
    out=file(filename, 'w')
    out.write('graph "SimpleGraph" {\n')
    out.write('overlap = "scale";\n')
    results = graph.query(query)
    donelinks = set()
    for binding in results:
        if binding[b1] != binding[b2]:
            n1, n2 = binding[b1].encode('utf-8'), binding[b2].encode('utf-8')
            if (n1, n2) not in donelinks and (n2, n1) not in donelinks:
                out.write('"%s" -- "%s"\n' % (n1, n2))
                donelinks.add((n1, n2))
    out.write('}')
```

This method queries the graph using the provided query, then loops over the resulting bindings, pulling out the variables b1 and b2 and creating a link between them. We can use this method to create a much cleaner celebrity dating graph:

```
>>> from simplegraph import *
>>> from graphtools import *
>>> cg = SimpleGraph()
>>> cg.load('celeb_triples.csv')
>>> querytodot(cg, [('?rel', 'with', '?p1'), ('?rel', 'with', '?p2')], 'p1', \
    'p2', 'relationships.dot')
>>> exit()
$ neato -Teps -Orelationships relationships.dot
```

A partial result is shown in Figure 3-7.

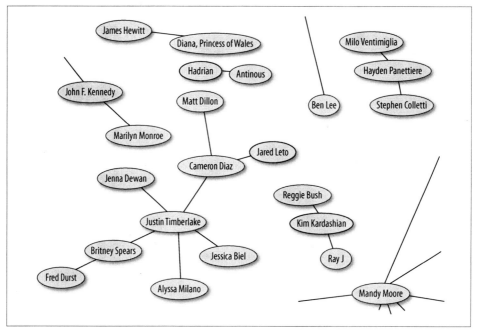

Figure 3-7. Viewing the celebrity dating graph

See if you can make other images from the business or movie data. For example, try graphing all the contributors to different politicians.

Semantic Data Is Flexible

An important point that we emphasize throughout this book is that semantic data is flexible. In Chapter 2 you saw how we could represent many different kinds of information using semantics, and in this chapter we've shown you some generic methods for querying and exploring *any* semantic database.

Now, let's say you were given a new set of triples, and you had no idea what it was. Using the techniques described in this chapter, you could immediately:

- Visualize the data to understand what's there and which predicates are used.
- Construct queries that search for patterns across multiple nodes.
- Search for connections between items in the graph.
- Build rules for inferring new information, such as geocoding of locations.
- Look for overlaps between this new data and an existing set of data, and merge the sets without needing to create a new schema.

You should now have a thorough grasp of what semantic data is, the domains it can work in, and what you can do once you represent data this way. In the following chapters we'll look at industry-standard representations and highly scalable implementations of semantic databases.

## PART II

Standards and Sources

## CHAPTER 4

Just Enough RDF

Now that you have an understanding of semantics and of how you might build a simple triplestore on your preferred platform, we're going to introduce you to the world of data formats and semantic standards. At this point, you may have a few unanswered questions after reading the previous chapter:

- How do I know which predicates (verbs) to use?
- How do I know whether a value in my data refers to a floating-point number or a string representing the number?
- Why is "San Francisco, CA" represented as San_Francisco_California and not something else?
- How will other people know what representations I used?
- Are comma-separated triples really the best way to store and share data?

These questions are concerned with *maintaining* and *sharing* a large set of semantic data. In this chapter, we'll introduce the following few concepts to help solve these problems:

- URIs and strong keys, so you can be sure you're talking about the same thing as someone else
- RDF serializations, which have parsers for every popular language and are the standard way to represent and share semantic data
- The SPARQL language, the standard way of querying semantic data

The use of semantic web formats goes far beyond what we can cover here, so consider these concepts as the highlights—the pieces you really need in order to get started building semantic applications.

What Is RDF?

Hopefully, by this point you've seen that structuring data into graphs is easy, and a good idea as well. But now that you've got your data loaded into a graph, how do you share it and make it available to other people? This is where the W3C's Resource Description Framework (RDF) comes into play. RDF provides a standard way of expressing graphs of data and sharing them with other people and, perhaps more importantly, with machines. Because it is a W3C "Recommendation" (an industry standard by any other measure), a large collection of tools and services has emerged around RDF. This chapter introduces just enough of the RDF standards to allow you to take advantage of these tools while avoiding much of the complexity. We'll be using the Python RDFLib library, but much like the variations in DOM APIs, you should be able to use what you learn in this chapter with any RDF API, as the principles remain the same across libraries.

The history of RDF goes back to 1990, when Tim Berners-Lee wrote a proposal that led to the development of the World Wide Web. In the original proposal, there were different types of links between documents, which made the hypertext easier for computers to comprehend automatically. Typed links were not included in the first HTML spec, but the ideas resurfaced in the Meta Content Framework (MCF), a system for describing metadata and organizing the Web that was developed by Ramanathan Guha while he was at Apple and Netscape in the late 1990s, with an XML representation developed with Tim Bray. The W3C was looking for a generic metadata representation, and many of the ideas in MCF found their way into the first RDF W3C Recommendation in 1999. Since then, the standards have been revised, and today's software and tools reflect these improvements.

The RDF Data Model

RDF is a language for expressing data models using statements expressed as triples. Just as we've seen in previous chapters, each statement is composed of a subject, a predicate, and an object. RDF adds several important concepts that make these models much more precise and robust. These additions play an important role in removing ambiguity when transmitting semantic data between machines that may have no other knowledge of one another.

URIs As Strong Keys

You may recall from Chapter 1 that when we wanted to join multiple tables of relational data together, we needed to find a shared identifier between the two tables that allowed us to map the tables together. These shared identifiers are called *keys* and form the basis for joining relational data.

Similarly, in a graph data structure, we need to assign unique identifiers to each of the nodes so that we can refer to them consistently across all the triples that describe their relationships. Up to this point, we have been using strings to label the nodes in our graphs, and we have been careful to use unique names whenever there was a chance of overlap. For instance, we used a clever naming convention to disambiguate different places with the same name in the `places_triples` dataset, in order that IDs such as `San_Francisco_California` versus `San_Francisco_Chile` would be distinct.

But for larger applications and when we are coordinating multiple datasets, it can become increasingly difficult to guarantee unique and consistent identifiers for each node. An example of this would be if we tried to merge a database of old classic films with our film database from Chapter 2. The classic film database would likely have an entry for Harrison Ford, the handsome star of silent classics such as *Oh, Lady, Lady*. If the ID for this actor was `harrison_ford`, then when we merged the two databases, we might sadly discover that the actor who played Han Solo and Indiana Jones was killed as the result of a car accident in 1957.

Resources

To avoid these types of ambiguities, RDF conceptualizes anything (and everything) in the universe as a resource. A *resource* is simply anything that can be identified with a Universal Resource Identifier (URI). And by design, anything we can talk about can be assigned a URI. You are probably most familiar with URLs (Universal Resource Locators), or the strings used to specify how web pages (among other things) are retrieved. URLs are a subset of URIs that identify where digital information can be retrieved. While URLs tell you where to find specific information, they also provide a unique identifier for the information. URIs generalize this concept further by saying that anything, whether you can retrieve it electronically or not, can be uniquely identified in a similar way. See Figure 4-1.

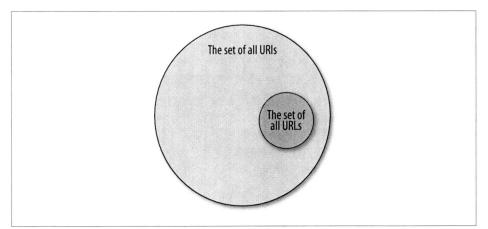

Figure 4-1. Venn diagram showing the relationship of URLs to URIs; URLs are a subset of URIs

Since URIs can identify anything as a resource, the subject of an RDF statement can be a resource, the object in an RDF statement can be a resource, and predicates in RDF statements are always resources. The URI for a resource represented in an RDF statement is called the *URI reference* (abbreviated URIref) for that graph node. The subtlety of using the phrase "URI reference" is easy to miss. That is, the node in the graph isn't the thing the URI identifies; rather, the URI is the identifier for something that is itself being represented as a node in a graph. Thus we say the graph node has a reference, which is the thing identified by the URI.

A Real-World URIref

Alongside Monet's first painting of Rouen Cathedral is a note from the curator of the collection that says, "This is Rouen Cathedral." Of course, it isn't the cathedral—it is a painting of the cathedral. You would think they would know the difference!

URIs are a useful way of getting around the need for an omniscient data architect. By allowing distributed organizations to create names for the things they are interested in, we can avoid the problem of two groups choosing the same name for different entities. URIs are simply strings, composed of a scheme name followed by a colon, two slashes (://), and a scheme-specific identifier. The scheme identifies the protocol for the URI, while the scheme-specific identifier is used by the protocol of the scheme to uniquely identify the resource. Almost all the URIs we will encounter use the "http" or "https" scheme for identifying things. As we have experienced with URLs, the scheme-specific part of these identifiers typically takes the form of a hostname, an optional port identifier, and a hierarchical path. If organizations stick to creating identifiers using hostnames under their control, then there is no chance of two organizations constructing the same identifier for two different things. Similarly, organizations themselves can split up their own identifier creation efforts among different hostnames and/or different hierarchical paths, thus making distributed resource naming work at any scale.

Because URIs uniquely identify resources (things in the world), we consider them *strong identifiers*. There is no ambiguity about what they represent, and they always represent the same thing, regardless of the context we find them in.

It is important to note that URIs are not URLs (although every URL is a URI). Practically speaking, this means that you shouldn't assume URIs will produce any information if you type the identifier into a web browser. That said, making digital information about the resource available at that address is considered good practice, as we will see in the section "Linked Data" on page 105.

It is common in RDF to shorten URIs by assigning a namespace to the base URI and writing only the distinctive part of the identifier. For instance, `rdf` is frequently used as a moniker for the base URI `http://www.w3.org/1999/02/22-rdf-syntax-ns#`, allowing predicates such as `http://www.w3.org/1999/02/22-rdf-syntax-ns#type` to be abbreviated as `rdf:type`.

Blank Nodes

You may have noticed that in our discussion about URIrefs and resources, we hedged our language and didn't say, "All RDF subjects are resources." That's because you may find situations where you don't know the URI of the thing you would like to reference or where there is no identifier available (and you are not in a good position to construct one). In either case, just because we don't have a URI for the item doesn't mean we can't talk about it. For these situations, RDF provides "anonymous" or blank nodes.

Blank nodes are graph nodes that represent a subject (or object) for which we would like to make assertions, but have no way to address with a proper URI. For instance, as we will see in Chapter 5, many social network APIs don't issue strong URIs for the members of their community, even though they have a good deal to say about them. Instead, the social network uses a blank node to represent the member, and the facts about that member are connected to the blank node. See Figure 4-2.

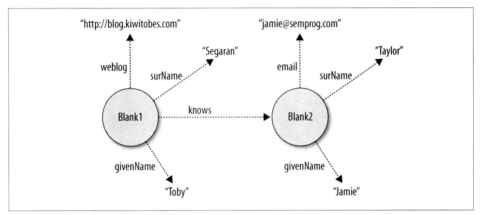

Figure 4-2. Blank nodes in a social graph

In triple representations, blank node IDs are written _:id, where id is an arbitrary, graph-specific local identifier. Most RDF APIs handle this by issuing an internal ID for the node that is only valid in the local graph and can't be used as a strong key between graphs. Using our previous triple expression format, we can write the graph in Figure 4-2 as:

```
(_:ax1, "weblog", "http://blog.kiwitobes.com")
(_:ax1, "surName", "Segaran")
(_:ax1, "givenName", "Toby")
(_:ax1, "knows", _:zb7)
(_:zb7, "surName", "Taylor")
(_:zb7, "givenName", "Jamie")
(_:zb7, "email", "jamie@semprog.com")
```

And though neither of the nodes has a strong external identifier, from the context of data connected to the nodes it is clear that the Toby Segaran who writes the Kiwitobes blog knows the Jamie Taylor with the email address *jamie@semprog.com*.

You may also find situations where it is useful to create a data model that uses a node in an RDF graph to simply group a set of statements together. For instance, while describing a building, it is often helpful to group together its street address, municipality, administrative region, and postal code as a unit that represents the building's mailing address. To do this, we would need a graph node, which represents the abstract concept of the building's address. While we could provide a URIref for this mailing address, we would almost never need to refer to this grouping node outside the context of the current graph. See Figure 4-3.

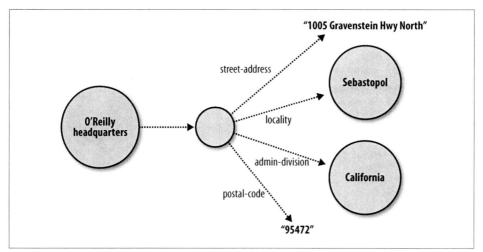

Figure 4-3. Using a blank node to model a mailing address

Literal Values

RDF uses literal values in the same way our earlier graph examples did, to express names, dates, and other types of values about a subject. In RDF, a literal value can optionally have a language (e.g., English, Japanese) or a type (e.g., integer, boolean, string) associated with it. Type URIs from the XML schema spec are commonly used to indicate literal types; for example, http://www.w3.org/2001/XMLSchema#int or xsd:int for an integer value. ISO 639 codes are used to specify the language; for example, en for English or ja for Japanese.

RDF Serialization Formats

While the data model that RDF uses is very simple, the serialized representation tends to get complicated when an RDF graph is saved to a file or sent over a network because of the various methods used to compact the data while still leaving it readable. These compaction mechanisms generally take the form of shortcuts that identify multiple references to a graph node using a shared but complex structure.

The good news is that you really don't have to worry about the complexities of the serialization formats, as there are open source RDF libraries for just about every modern programming language that handle them for you. (We believe the corollary is also true:

if you are thinking about the serialization format and you aren't in the business of writing an RDF library, then you should probably find a better library.) Because of this we won't go into too much detail, but it is important to understand the basics so you can use the most appropriate format and debug the output.

We'll be covering four serialization formats here: N-Triples, the simplest of notations; N3, a compaction of the N-Triple format; RDF/XML, one of the most frequently used serialization formats; and finally, "RDF in attributes" (known as RDFa), which can be embedded in other serialization formats such as XHTML.

 There are many books and online resources that cover these output formats in great detail. If you are interested in reading further about them, you can look at the complete RDF specification at *http://www.w3.org/RDF/* or in the O'Reilly book *Practical RDF* by Shelley Powers.

A Graph of Friends

In order to compare the different serialization formats, let's first build a simple graph that we can use throughout the examples to observe how the various serializations fold relationships together.

For this example graph, we'll model a small part of Toby's social sphere—in particular, how he knows the other authors of this book. In our graph we will not only include information about the people Toby knows, but we'll also describe other relationships that can be used to uniquely identify Toby. This will include things like the home page of his blog, his email address, his interests, and any other names he might use.

These clues about Toby's identity are important to help differentiate "our Toby" from the numerous other Tobys in the world. Human names are hardly unique, but by providing a collection of attributes about the person, hopefully we can pinpoint the individual of interest and obtain a strong identifier (URI) that we can use for future interaction.

As you might have discerned, the network of social relationships that people have with one another naturally lends itself to a graphical representation. So it is probably no surprise that machine-readable graphs of friends have coevolved with RDF, making social graphs one of the most widely available RDF datasets on the public Internet. Over time, the relationships expressed in these social graphs have settled into a collection of well-known predicates, forming a vocabulary of expression known as "Friend of a Friend" or simply FOAF.

Not surprisingly, the core FOAF vocabulary—the set of predicates—has been adopted and extended to describe a number of common "things" available on the Internet. For instance, FOAF provides a predicate for identifying photographs that portray the subject of the statement. While FOAF deals primarily with people, the formal definition of FOAF states that the `foaf:depiction` predicate can be used for graphics portraying any resource (or "thing") in the world.

Figure 4-4 represents the small slice of Toby's social world that we will concern ourselves with over the next few examples. With this graph in mind, let's look at how this knowledge can be represented using different notations.

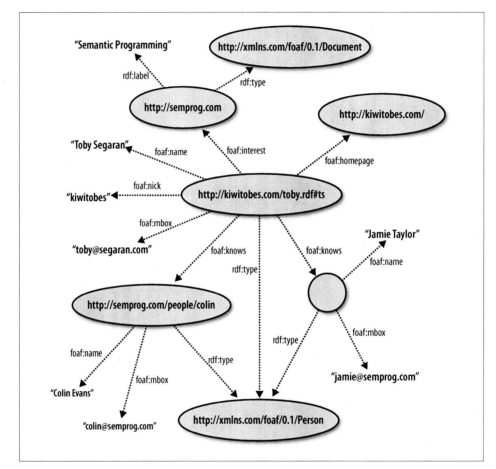

Figure 4-4. Toby's FOAF graph

N-Triples

N-Triple notation is a very simple but verbose serialization, similar to what we have been using in our triple data files up to this point. Because of their simplicity, N-Triples were used by the W3C Core Working Group to unambiguously express various RDF test-case data models while developing the updated RDF specification. This simplicity also makes the N-Triple format useful when hand-crafting datasets for application testing and debugging.

Each line of output in N-Triple format represents a single statement containing a subject, predicate, and object followed by a dot. Except for blank nodes and literals, subjects, predicates, and objects are expressed as absolute URIs enclosed in angle brackets. Subjects and objects representing anonymous nodes are represented as _:name, where name is an alphanumeric node name that starts with a letter. Object literals are double-quoted strings that use the backslash to escape double-quotes, tabs, newlines, and the backslash character itself. String literals in N-Triple notation can optionally specify their language when followed by @lang, where lang is an ISO 639 language code. Literals can also provide information about their datatype when followed by ^^type, where type is commonly an XSD (XML Schema Definition) datatype.

The extension *.nt* is typically used when N-Triples are stored in a file, and when they are transmitted over HTTP, the mime type text/plain is used. The official N-Triple format is documented at *http://www.w3.org/TR/rdf-testcases/#ntriples*.

Our FOAF graph (as shown in Figure 4-4) can be represented in N-Triple format as:

```
<http://kiwitobes.com/toby.rdf#ts> <http://xmlns.com/foaf/0.1/homepage>
    <http://kiwitobes.com/>.
<http://kiwitobes.com/toby.rdf#ts> <http://xmlns.com/foaf/0.1/nick> "kiwitobes".
<http://kiwitobes.com/toby.rdf#ts> <http://xmlns.com/foaf/0.1/name> "Toby Segaran".
<http://kiwitobes.com/toby.rdf#ts> <http://xmlns.com/foaf/0.1/mbox>
    <mailto:toby@segaran.com>.
<http://kiwitobes.com/toby.rdf#ts> <http://xmlns.com/foaf/0.1/interest>
    <http://semprog.com>.
<http://kiwitobes.com/toby.rdf#ts> <http://www.w3.org/1999/02/22-rdf-syntax-ns#type>
    <http://xmlns.com/foaf/0.1/Person>.

<http://kiwitobes.com/toby.rdf#ts> <http://xmlns.com/foaf/0.1/knows> _:jamie .
<http://kiwitobes.com/toby.rdf#ts> <http://xmlns.com/foaf/0.1/knows>
    <http://semprog.com/people/colin>.

_:jamie <http://xmlns.com/foaf/0.1/name> "Jamie Taylor".
_:jamie <http://xmlns.com/foaf/0.1/mbox> <mailto:jamie@semprog.com>.
_:jamie <http://www.w3.org/1999/02/22-rdf-syntax-ns#type>
    <http://xmlns.com/foaf/0.1/Person>.

<http://semprog.com/people/colin> <http://xmlns.com/foaf/0.1/name> "Colin Evans".
<http://semprog.com/people/colin> <http://xmlns.com/foaf/0.1/mbox>
    <mailto:colin@semprog.com>.
<http://semprog.com/people/colin> <http://www.w3.org/1999/02/22-rdf-syntax-ns#type>
    <http://xmlns.com/foaf/0.1/Person>.

<http://semprog.com> <http://www.w3.org/2000/01/rdf-schema#label>
    "Semantic Programming".
<http://semprog.com> <http://www.w3.org/1999/02/22-rdf-syntax-ns#type>
    <http://xmlns.com/foaf/0.1/Document>.
```

N3

While N-Triples are conceptually very simple, you may have noticed a lot of repetition in the output. The redundant information takes additional time to transmit and parse. While it's not a problem when working with small amounts of data, the additional information becomes a liability when working with large amounts of data. By adding a few additional structures, N3 condenses much of the repetition in the N-Triple format.

In an RDF graph, every connection between nodes represents a triple. Since each node may participate in a large number of relationships, we could significantly reduce the number of characters used in N-Triples if we used a short symbol to represent repeated nodes. We could go further, recognizing that many of the URIs used in a specific model frequently come from related URIs. In much the same way that XML provides a namespace mechanism for generating short Qualified Name (qnames) for nodes, N3 allows us to define a URI prefix and identify entity URIs relative to a set of prefixes declared at the beginning of the document. The statement:

```
@prefix semperp: <http://semprog.com/people/>.
```

allows us to shorten the absolute URI for Colin from `<http://semprog.com/people/colin>` to `semperp:colin`.

Since each node in an RDF graph is a potential subject about which we may have many things to say, it is not uncommon to see the same subject repeat many (many) times in N-Triple output. N3 reduces this repetition by allowing you to combine multiple statements about the same subject by using a semicolon (;) after the first statement, so you only need to state the predicate and object for other statements using the same subject. The following statement says that Colin knows Toby and that Colin's email address is *colin@semprog.com* (note how `semperp:colin`, the subject, is only stated once):

```
semperp:colin foaf:knows <http://kiwitobes.com/toby.rdf#ts>;
    foaf:mbox "colin@semprog.com".
```

N3 also provides a shortcut that allows you to express a group of statements that share a common anonymous subject (blank node) without having to specify an internal name for the blank node. As discussed earlier, mailing addresses are frequently modeled with a blank node to hold all the components of the address together. W3C has defined a vocabulary for representing the data elements of the vCard interchange format (*http://www.w3.org/2006/vcard/*) that includes predicates for modeling street addresses. For instance, to specify the address of O'Reilly, you could write:

```
[ <http://www.w3.org/2006/vcard/ns#street-address> "1005 Gravenstein Hwy North" ;
    <http://www.w3.org/2006/vcard/ns#locality> "Sebastopol, California"
].
```

Because it is important to explicitly state that an entity is of a certain type, N3 allows you to use the letter a as a predicate to represent the RDF "type" relationship represented by the URI `<http://www.w3.org/1999/02/22-rdf-syntax-ns#type>`.

Another predicate for which N3 provides a shortcut is <http://www.w3.org/2002/07/owl#sameAs>. OWL (Web Ontology Language) is a vocabulary for defining precise relationships between model elements. We will have more to say about OWL in Chapter 6, but even when models don't use the precision of OWL, you will frequently see the owl:sameAs predicate to express that two URIs refer to the same entity. The sameAs predicate is used so frequently that the authors of N3 designated the symbol = as shorthand for it.

Because N-Triples are a subset of N3, any library capable of reading N3 will also read N-Triples. The FOAF graph (Figure 4-4) in N3 would read:

```
@prefix foaf: <http://xmlns.com/foaf/0.1/>.
@prefix rdf: <http://www.w3.org/1999/02/22-rdf-syntax-ns#>.
@prefix rdfs: <http://www.w3.org/2000/01/rdf-schema#>.
@prefix semperp: <http://semprog.com/people/>.
@prefix tobes: <http://kiwitobes.com/toby.rdf#>.

tobes:ts a foaf:Person;
    foaf:homepage <http://kiwitobes.com/>;
    foaf:interest <http://semprog.com>;
    foaf:knows semperp:colin,
        [ a foaf:Person;
            foaf:mbox <mailto:jamie@semprog.com>;
            foaf:name "Jamie Taylor"];
    foaf:mbox <mailto:toby@segaran.com>;
    foaf:name "Toby Segaran";
    foaf:nick "kiwitobes".

<http://semprog.com> a foaf:Document;
    rdfs:label "Semantic Programming".

semperp:colin a foaf:Person;
    foaf:mbox <mailto:colin@semprog.com>;
    foaf:name "Colin Evans".
```

RDF/XML

The original W3C Recommendation on RDF covered both a description of RDF as a data model and XML as an expression of RDF models. Because of this, people sometimes refer to RDF/XML as RDF, but it is important to recognize that it is just one possible representation of an RDF graph. RDF/XML is sometimes criticized for being difficult to read due to all the abbreviated structures it provides; still, it is one of the most frequently used formats, so it's useful to have some familiarity with its layout.

Conceptually, RDF/XML is built up from a series of smaller descriptions, each of which traces a path through an RDF graph. These paths are described in terms of the nodes (subjects) and the links (predicates) that connect them to other nodes (objects). This sequence of "node, link, node" is repeatable, forming a "striped" structure (think of a candy cane, with nodes being red stripes and predicates being white stripes). Since each node encountered in these descriptions has a strong identifier, it is possible to weave the smaller descriptions together to learn the larger RDF graph structure. See Figure 4-5.

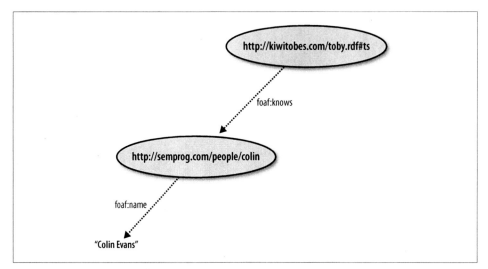

Figure 4-5. A stripe from Toby's FOAF graph

If there is more than one path described in an RDF/XML document, all the descriptions must be children of a single RDF element; if there is only one path described, the `rdf:RDF` element may be omitted. As with other XML documents, the top-level element is frequently used to define other XML namespaces used throughout the document:

```
<rdf:RDF xmlns:rdf="http://www.w3.org/1999/02/22-rdf-syntax-ns#"/>
```

Paths are always described starting with a graph node, using an `rdf:Description` element. The URI reference for the node can be specified in the description element with an `rdf:about` attribute. For blank nodes, a local identifier (valid only within the context of the current document) can be specified using an `rdf:NodeID` attribute. Predicate links are specified as child elements of the `rdf:Description` node, which will have their own children representing graph nodes. The simple stripe representing Colin as a friend of Toby (Figure 4-5) would look like:

```
<rdf:RDF xmlns:rdf="http://www.w3.org/1999/02/22-rdf-syntax-ns#"
         xmlns:foaf="http://xmlns.com/foaf/0.1/">

  <rdf:Description rdf:About="http://kiwitobes.com/toby.rdf#ts>
    <foaf:knows>
      <rdf:Description rdf:About="http://semprog.com/people/colin">
        <foaf:name>Colin Evans</foaf:name>
      </rdf:Description>
    </foaf:knows>
  </rdf:Description>

</rdf:RDF>
```

Literal objects can be specified as the text of an element, or as an attribute on the
rdf:Description element. Let's expand the example, adding more information about
Colin and about another friend of Toby's:

```
<rdf:RDF xmlns:rdf="http://www.w3.org/1999/02/22-rdf-syntax-ns#"
         xmlns:foaf="http://xmlns.com/foaf/0.1/">

   <rdf:Description rdf:about="http://kiwitobes.com/toby.rdf#ts">
     <foaf:knows>
       <rdf:Description rdf:about="http://semprog.com/people/colin">
         <foaf:name>Colin Evans</foaf:name>
         <foaf:mbox>colin@semprog.com</foaf:mbox>
       </rdf:Description>
     </foaf:knows>

     <foaf:knows>
       <rdf:Description foaf:mbox="jamie@semprog.com"/>
     </foaf:knows>
   </rdf:Description>

</rdf:RDF>
```

While this is a perfectly reasonable description of Toby's relationship to Colin and
Jamie, we are still missing the rdf:type information that states that Toby, Colin, and
Jamie are people. As in the other RDF serializations we have looked at, RDF/XML
provides a shortcut for this very common statement, allowing you to replace the
rdf:Description element with an element representing the rdf:type for the node. Thus
the sequence of elements:

```
<rdf:Description rdf:about="http://www.kiwitobes.com/toby.rdf#ts"><rdf:type>
    <foaf:Person>
```

is compacted into a single rdf:Description element of the form:

```
<foaf:Person rdf:about="http://kiwitobes.com/toby.rdf#ts">
```

The FOAF graph we represented in N-Triples and N3 can now be represented in RDF/
XML as:

```
<rdf:RDF
  xmlns:foaf='http://xmlns.com/foaf/0.1/'
  xmlns:rdf='http://www.w3.org/1999/02/22-rdf-syntax-ns#'
  xmlns:rdfs='http://www.w3.org/2000/01/rdf-schema#'>

  <foaf:Person rdf:about="http://kiwitobes.com/toby.rdf#ts">
    <foaf:name>Toby Segaran</foaf:name>
    <foaf:homepage rdf:resource="http://kiwitobes.com/"/>
    <foaf:nick>kiwitobes</foaf:nick>
    <foaf:mbox rdf:resource="mailto:toby@segaran.com"/>

    <foaf:interest>
      <foaf:Document rdf:about="http://semprog.com">
        <rdfs:label>Semantic Programming</rdfs:label>
      </foaf:Document>
    </foaf:interest>
```

```
  <foaf:knows>
    <foaf:Person rdf:about="http://semprog.com/people/colin">
      <foaf:name>Colin Evans</foaf:name>
      <foaf:mbox rdf:resource="mailto:colin@semprog.com"/>
    </foaf:Person>

  </foaf:knows>
  <foaf:knows>
    <foaf:Person>
      <foaf:name>Jamie Taylor</foaf:name>
      <foaf:mbox rdf:resource="mailto:jamie@semprog.com"/>
    </foaf:Person>
  </foaf:knows>

</foaf:Person>
</rdf:RDF>
```

These aren't the only abbreviated structures RDF/XML provides, but this should be enough to let you read most RDF/XML files.

RDFa

RDFa isn't a pure serialization format for RDF, but rather a way of annotating XHTML web pages with RDF data. The idea behind RDFa is that you only have to publish your content once, mixing the human-readable and machine-readable content together. This is a similar philosophy to that of Microformats, a simpler, more ad-hoc approach to adding rich semantic annotations to XHTML content.

RDFa uses a small set of XML attributes that are added to existing XHTML content tags in order to specify the semantics behind the information that is displayed. These attributes make the semantic meaning of existing XHTML content clear. The basic processing model is that the subject of a triple is the subject URI identified in a higher-level XHTML element in the DOM tree, and the predicate and object of a statement are lower down on the tree, children of the subject.

Instead of using URIs to describe subjects, predicates, and objects, many RDFa attributes use Compact URIs (or CURIEs) to reduce the amount of markup. CURIEs work just like XML Qualified Names (in fact, QNames are a subset of CURIEs), so everything you know about XML QNames (such as that `foaf:nick` actually means `http://xmlns.com/foaf/0.1/nick`) applies to CURIEs. But CURIEs are a bit more accommodating in what the `localpart` of the `prefix:localpart` expression can contain.

QName construction forbids slashes (/) in the `localpart`, thus requiring a separate XML namespace declaration for every QName using a different part of the path hierarchy. CURIEs relax this constraint, allowing statements like `example:cow` and `example:places/barn` to use one xmlns declaration—like `http://example.org/farm/`— to generate the full URIs `http://example.org/farm/cow` and `http://example.org/farm/places/barn`, respectively.

CUIREs also allow for `localparts` that start with a number. This means that you could define an XML namespace as:

```
xmlns:amazonisbn="http://www.amazon.com/exec/obidos/ASIN/"
```

and then refer to the book *Programming the Semantic Web* by its ISBN with the CURIE:

```
amazonisbn:0596153813
```

CURIEs are great when you are working with predicates because you can make one xmlns declaration for each vocabulary you are using and quickly construct CURIEs for any property in the vocabulary. However, they can be frustrating when you want to talk about a wide range of subjects or objects (since you have to make an xmlns declaration for each unique base URI). To alleviate this problem, RDFa allows you to use full URIs for several of the subject and object attributes. But because full URIs use a colon to separate the protocol scheme from the hierarchical part of the URI, parsers could become confused when they see the `http:`—did you mean `http:` as in `http://example.org/cow` or were you writing a CURIE where `http` is a prefix for some namespace?

To avoid this confusion, RDFa defines a "safe CURIE" that makes it clear when a colon-separated statement is being used as a CURIE versus as a protocol identifier in a URI. To construct a safe CURIE, simply place your CURIE in square brackets, as in:

```
[example:place/barn]
```

Let's look at the list of attributes used by RDFa, grouping them by the part of the RDF statement that they declare.

This is the attribute to set an RDF subject:

about
> A URI (or safe CURIE) used as a subject in an RDF triple. By default, the base URI for the page is the root URI for all statements. Using an `about` attribute allows statements to be made where the base URI isn't the subject.

These are the attributes to set an RDF predicate:

rel
> CURIEs expressing relationships between two resources

property
> CURIEs expressing relationships between a resource and a literal

rev
> CURIEs expressing a reverse relationship between two resources

These are the attributes to set an RDF object:

content
> A string, representing a literal RDF object

href

    A URI resource expressing an RDF object (as inline clickable)

src

    A URI resource expressing an RDF object (as an inline embedded item)

resource

    A URI (or safe CURIE) expressing an RDF object when the object isn't visible on the page

RDFa also provides special attributes for specifying datatypes and making `rdf:type` statements:

datatype

    The datatype of a literal

typeof

    The type of a subject

It is important to note that the attribute you use to set the predicate depends on the type of object in the RDF statement. If the object is a literal, then the predicate is specified with the `property` attribute. If, however, the object is a resource, the `rel` or `rev` attribute is used. Because XHTML is markup for producing human-readable displays, it may not be convenient to display data in the subject-predicate-object order of an RDF statement. To handle these situations, RDFa provides the `rev` attribute for setting the predicate, which also indicates that the order of the statement has been reversed (object, predicate, subject).

Minimally, we can specify RDF triples in a single markup element. In the following example, the object is a literal, so we use the attribute `property` to state the predicate:

```
<span xmlns:foaf="http://xmlns.com/foaf/0.1/"
      about="http://kiwitobes.com/toby.rdf#ts"
      property="foaf:nick"
      content="kiwitobes" />
```

When the statement's object is a resource, we use the attribute `rel` to state the predicate:

```
<span xmlns:foaf="http://xmlns.com/foaf/0.1/"
      about="http://kiwitobes.com/toby.rdf#ts"
      rel="http://xmlns.com/foaf/0.1/homepage"
      href="http://kiwkitobes.com" />
```

But we can also use the XHTML element to display parts of the structure. For instance, we can make the following statement about Toby's nickname when displaying the string "kiwitobes":

```
Toby's nickname is: <span xmlns:foaf="http://xmlns.com/foaf/0.1/"
      about="http://kiwitobes.com/toby.rdf#ts"
      property="http://xmlns.com/foaf/0.1/nick">kiwitobes</span>
```

Here's an example of Toby's FOAF record as a fragment of XHTML annotated with RDFa. The XML attributes and text that an RDFa parser would glean from this

XHTML are in bold. The annotations can be made in any tags and are meant to reuse existing XHTML attributes and text that are also being used in the human-readable XHTML content. It can be tricky to figure out how to weave the attributes into existing documents, but the benefit is that all of the data is made available in one place and the concepts being discussed are unambiguously described using strong URIrefs:

```
<div xmlns:foaf="http://xmlns.com/foaf/0.1/"
      about="http://kiwitobes.com/toby.rdf#ts" typeof="foaf:Person">

  Name: <span property="foaf:name">Toby Segaran</span><br/>
  Nickname: <span property="foaf:nick">kiwitobes</span><br/>
  Interests: <a rel="foaf:interest" href="http://semprog.org">
               <span property="rdfs:label">Semantic Programming</span></a>
  Homepage: <a rel="foaf:homepage" href="http://kiwkitobes.com/">KiwiTobes</a><p/>


  Friends:<br/>
  <ul rel="foaf:knows">
    <li about="http://semprog.com/people/colin"
        typeof="foaf:Person" property="foaf:name">Colin Evans</li>

    <li typeof="foaf:Person">
      <span property="foaf:name">Jamie Taylor</span><br/>
      Email: <a rel="foaf:mbox" href="mailto:jamie@semprog.com">
          jamie@semprog.com</a><br/>
    </li>

  </ul>
</div>
```

 In documents with copious amounts of human markup, it can be challenging to read RDFa. One way to work your way through the jungle of markup is to scan for rel, rev, and property attributes in a markup tag. Once you have found one of these elements, you know you have found the predicate of a statement. Then, search backward up the DOM tree to find the subject of the statement (remember, if you don't find one, the document itself is the subject). Then search down the DOM tree to find the next item that can serve as an object for the statement. Keep in mind that if the predicate was specified using a rev attribute, the order of the statement will be reversed.

Because XHTML was designed to be extensible and allows new attributes to be added to markup elements, RDFa was specified as annotations on XHML. In practice, RDFa works perfectly well on HTML, though the markup will not validate against HTML 4.

Because it is easy to wrap templated HTML output with RDFa, and given the prevalence of database-driven websites, the amount of RDFa available on the Web is growing rapidly. A number of prominent sites, including MySpace and popular authoring tools, now have RDFa output capabilities, though it may not be obvious because RDFa doesn't alter the HTML rendering.

In later chapters we will show how Yahoo! is extracting semantic data from sites using RDFa to enhance search, and we will revisit RDFa as we build more sophisticated semantic applications that not only consume semantic data, but republish their output for use by other semantic services.

RDFa is now a W3C Recommendation, but there have been several other attempts to embed RDF in HTML. One effort was eRDF ("embeddable RDF"), which predates RDFa. eRDF never reached a critical mass, and like other RDF microformats, it generally isn't supported by RDF tools; still, it's possible that you may run into it. You can read more about eRDF at *http://research.talis.com/2005/erdf/wiki/Main/RdfInHtml*.

Not the Last Word on Serialization

N-Triples, N3, RDF/XML, and RFDa are not the only RDF serializations you will find in the wild. Turtle is another popular and fairly simple serialization with its own compaction tricks. Turtle output is typically associated with the mime type `application/x-turtle` and the file extension *.ttl*. While you can learn more about it at *http://www.dajobe.org/2004/01/turtle/*, we believe that you need only be aware of its existence and know how to tell your favorite RDF library to read it.

If you run into a serialization that you find difficult to read or debug, try reading the data into your RDF library and then asking the library to serialize the graph back into a format you are comfortable with.

Introducing RDFLib

The triplestore and query language we developed in Chapters 2 and 3 were useful for understanding how triplestores work. And while we could implement parsers and serializers for the myriad RDF output formats and add all the features you would expect in a full-fledged RDF library, doing so would distract us from our real interest: building semantic applications for the Web.

The good news is that research on semantic web technologies over the past decade has produced a number of excellent open source libraries for managing RDF data. For the remainder of this book we will adopt various RDF libraries and frameworks, introducing you to some of the more popular semantic platforms and allowing us to select the tool best suited for a particular task.

For many of our examples we will utilize RDFLib, a lightweight but functionally complete RDF library. RDFLib is very Pythonic in its approach, allowing applications to access RDF structures through standard Python idioms. You can download RDFLib from *http://rdflib.net*. To install the library, decompress and unpack the TAR file and make the top-level directory of the project your working directory. At a command prompt, enter:

```
c:\download\rdflib-2.4.0>python setup.py install
```

Now that you have RDFLib installed, let's take it for a test drive and see what we can do with it. Start an interactive Python session and create a graph with the following commands:

```
>> import rdflib
>> from rdflib.Graph import ConjunctiveGraph
>> g = ConjunctiveGraph()
```

Let's read in an RDF graph from the Web and examine a few of the ways you can inspect it. The Graph class provides a handy parse method, which not only loads data files into the graph, but can retrieve them from the Web as well. Grab Colin's FOAF file from the semprog website with the following command:

```
>> g.parse("http://semprog.com/people/colin", format="nt")
```

You can view the triples that define the graph by iterating over the graph directly:

```
>> for triple in g:
>> ...    print triple
```

Or you can query the triples, just as we did in Chapter 2, by looking for triple patterns using wildcards within a statement:

```
>> list( g.triples((None,rdflib.URIRef('foaf:knows'),None)) )
```

Colin's FOAF file was originally in N-Triple format, but perhaps we would like to save it to disk as RDF/XML:

```
>> outfile = open("colin.xml", "w")
>> outfile.write(g.serialize(format="pretty-xml"))
```

If you open *colin.xml* in a text editor, you should see a FOAF file similar to the one we examined earlier in the section "RDF/XML" on page 73. Let's create a new graph and read Colin's FOAF file in from disk, and then look at the graph serialized as N3:

```
>> newg = ConjunctiveGraph()
>> newg.parse("colin.xml")
>> newg.serialize(format="n3")
```

We now have two graphs in memory, g and newg. If everything is working as expected, both graphs should be identical—the various serializations should not have changed any of the information in the graphs. To prove this to ourselves, we can make use of RDFLib's ability to perform set operations on graphs. To do this, we will subtract graph g from newg, which will remove all the triples that appear in g from newg. If newg and g are identical, we shouldn't have any triples in newg after subtracting g:

```
>> newg -= g
>> len(newg)
0
```

Actually, RDFLib provides a Graph method that tests whether two graphs have the same shape (in other words, are isomorphic). Let's try that again:

```
>> g.isomorphic(newg)
False
```

```
#we made newg have zero triples in our last test, so let's reload it and try again
>> newg.parse("colin.xml")
>> g.isomorphic(newg)
True
```

Just like our triplestore in Chapter 2, triples can be inserted directly into the graph using tuples. Since we believe that anyone who reads this book must be a friend, let's add you to Colin's social network. First, we need to add a statement that says you are a person (you are a person, right?). To do this, you will need to identify yourself with a URI reference (you can just make up a URI for now). RDFLib provides a class for creating URI references called URIRef that takes a string, representing the URI, as an argument:

```
>> me = URIRef("http://my.uri.com/goes/here")
```

Next, we need to create a URIref for the predicate `rdf:type`. Since we may need to add other URIrefs to the RDF vocabulary, we will use RDFLib's `Namespace` class to generate this URIref. Once instantiated, the `Namespace` class allows you to create URIrefs by accessing the instance as a dictionary:

```
>> RDF = rdflib.Namespace("http://www.w3.org/TR/rdf-schema/#")
>> rdf-type-predicate = RDF["type"]
```

Since we are adding new statements to Colin's FOAF graph, which is currently in memory and bound to the variable g, the FOAF namespace must have already been defined. To find out, we can list all the namespace bindings in the current graph:

```
>> [ x for x in g.namespaces() ]
[(u'foaf', rdflib.URIRef('http://xmlns.com/foaf/0.1/')),....]
```

The base URI for the FOAF vocabulary has been bound to the `foaf` namespace prefix; knowing that, we can go ahead and assert our statement, declaring you are a person:

```
>> g.add((me, rdf-type-predicate, foaf["person"]))
```

Using these same techniques, we can now add a statement declaring you are friend of Colin:

```
>> g.add(URIRef("http://semprog.com/people/colin"), foaf["knows"], me)
```

And like our triplestore in Chapter 2, the graph can be queried using tuples where None indicates a free parameter:

```
>> list( g.triples((None, foaf["knows"], None)) )
```

While RDFLib can use Python strings as subjects, predicates, and objects, some operations will not work if they are not properly typed as `rdflib.URIRef`. Since it is so easy to create URIrefs with RDFLib, and we know that predicates, subjects, and nonliteral objects should always have URI references, there isn't any reason for not constructing them correctly.

Persistence with RDFLib

One of the advantages of using RDFLib is that we can load triples into a graph store that will persist across our example applications and command-line sessions. With a persistent graph, we can then write code that will attach to the graph store and make use of the previously loaded triples.

RDFLib uses a simple plug-in framework that facilitates adding new parsers, serializers, and storage systems. While RDFLib supports the use of MySQL and Sleepycat databases for persistence, we will use SQLite as our storage medium to simplify setup. SQLite is a simple, cross-platform database system that runs within an application process and requires no configuration—perfect for our needs. If your system doesn't have SQLite installed, download and install a SQLite package from *http://www.sqlite.org/*.

The Python interface for SQLite is called pysqlite and can be downloaded from *http://pysqlite.org* (which should redirect you to the current home of the project). For Windows, download and run the binary installer appropriate for your version of Python. For all other platforms, download the compressed source TAR file and enter the following at the command prompt (replacing the version number to reflect the file you downloaded):

```
$ gunzip pysqlite-2.5.0a.tar.gz
$ tar xvf pysqlite-2.5.0a.tar
$ cd pysqlite-2.5.0
$ python setup.py build
$ python setup.py install
```

To initialize a new persistent RDFLib triplestore called `rdf-test.ts`, try the following from an interactive Python session:

```
>> import rdflib
>> from rdflib import Literal
>> store = rdflib.plugin.get('SQLite', rdflib.store.Store)('rdf-test.ts')
>> store.open('.', create=True) #create in the current working directory - dot
>> g = rdflib.ConjunctiveGraph(store)
>> semprog = rdflib.Namespace("http://semprog.com/people/")
>> foaf = rdflib.Namespace("http://xmlns.com/foaf/0.1/")
>> g.add((semprog["jamie"], foaf["name"], Literal("Jamie Taylor")))
>> g.add((semprog["jamie"], foaf["mbox"], Literal("jamie@semprog.com")))
>> g.serialize(format="nt") #just to check our work
>> g.commit()
```

Now quit your Python session (just to prove to yourself that there is no magic happening in your session memory). Then restart your Python interpreter and try reading the triples back from the persistent store:

```
>> import rdflib
>> store = rdflib.plugin.get('SQLite', rdflib.store.Store)('rdf-test.ts')
>> store.open('.', create=False)
>> g = rdflib.ConjunctiveGraph(store)
>> g.serialize(format="nt")
```

You should see the triples you entered during the first session. We should point out a few things before we move along. First, the default for the `create` parameter in the `open` method is `create=True`, so when you want to open an existing database, you must explicitly state `create=False`. Fortunately, it is an error to create a database that already exists, so if you forget and try to recreate the database, you will throw a pysqlite `OperationalError`, which will prevent the database from being overwritten. Also, pysqlite supports transactions, which RDFLib will use when writing to the store. By default, autocommit is off, which means you must explicitly call commit at the end of your writes.

Armed with the information from this introduction, you should now be able to navigate RDFLib's online documentation to find other useful classes and methods if you need to solve more complex problems.

SPARQL

Just as SQL provides a (relatively) standard query language across relational database systems, SPARQL provides a standardized query language for RDF graphs. SPARQL (Simple Protocol and RDF Query Language) is similar to the query language we developed in Chapter 3, with a number of important and powerful additions, including the ability to filter results and construct new graphs based on queries. Like our earlier query language, SPARQL queries attempt to match patterns in the graph and bind wildcard variables as it finds solutions.

Throughout this section we will query a small graph of movie data, represented as follows using N3. This set of statements comes from a larger set of triples derived from Freebase that represents movies released between the years 2000 and 2008. You can download both datasets from *http://semprog.com/psw/chapt4/moviedata*:

```
@prefix fb: <http://rdf.freebase.com/ns/> .

<http://rdf.freebase.com/ns/en.hollywood_homicide>
    <http://rdf.freebase.com/ns/film.film.directed_by>
    <http://rdf.freebase.com/ns/en.ron_shelton> ;
<http://rdf.freebase.com/ns/film.film.starring>
    <http://rdf.freebase.com/ns/en.harrison_ford> ,
    <http://rdf.freebase.com/ns/en.kurupt> ,
    <http://rdf.freebase.com/ns/en.robert_wagner> ;
<http://rdf.freebase.com/ns/film.film.initial_release_date> "2003" .

<http://rdf.freebase.com/ns/en.k_19_the_widowmaker>
    <http://rdf.freebase.com/ns/film.film.directed_by>
    <http://rdf.freebase.com/ns/en.kathryn_bigelow> ;
<http://rdf.freebase.com/ns/film.film.starring>
    <http://rdf.freebase.com/ns/en.harrison_ford> ,
    <http://rdf.freebase.com/ns/en.joss_ackland> ;
<http://rdf.freebase.com/ns/film.film.initial_release_date> "2002" .
```

```
<http://rdf.freebase.com/ns/en.dark_blue>
    <http://rdf.freebase.com/ns/film.film.directed_by>
    <http://rdf.freebase.com/ns/en.ron_shelton> ;
<http://rdf.freebase.com/ns/film.film.starring>
    <http://rdf.freebase.com/ns/en.kurupt> ,
    <http://rdf.freebase.com/ns/en.kurt_russell> ,
    <http://rdf.freebase.com/ns/en.dash_mihok> .

<http://rdf.freebase.com/ns/en.the_weight_of_water_2002>
    <http://rdf.freebase.com/ns/film.film.directed_by>
    <http://rdf.freebase.com/ns/en.kathryn_bigelow> ;
<http://rdf.freebase.com/ns/film.film.starring>
    <http://rdf.freebase.com/ns/en.sean_penn> ,
    <http://rdf.freebase.com/ns/en.elizabeth_hurley> ;
<http://rdf.freebase.com/ns/film.film.initial_release_date> "2002" .

<http://rdf.freebase.com/ns/en.becoming_dick>
    <http://rdf.freebase.com/ns/film.film.directed_by>
    <http://rdf.freebase.com/ns/en.bob_saget> ;
<http://rdf.freebase.com/ns/film.film.starring>
    <http://rdf.freebase.com/ns/en.robert_wagner> ,
    <http://rdf.freebase.com/ns/en.bob_saget> ;
<http://rdf.freebase.com/ns/film.film.initial_release_date> "2000" .

<http://rdf.freebase.com/ns/en.body_of_lies>
    <http://rdf.freebase.com/ns/film.film.directed_by>
    <http://rdf.freebase.com/ns/en.ridley_scott> ;
<http://rdf.freebase.com/ns/film.film.starring>
    <http://rdf.freebase.com/ns/en.russell_crowe> ,
    <http://rdf.freebase.com/ns/en.mark_strong> ;
<http://rdf.freebase.com/ns/film.film.initial_release_date> "2008" .

<http://rdf.freebase.com/ns/en.kurt_russell>
    <http://rdf.freebase.com/ns/type.object.name> "Kurt Russell" .
<http://rdf.freebase.com/ns/en.dash_mihok>
    <http://rdf.freebase.com/ns/type.object.name> "Dash Mihok" .
<http://rdf.freebase.com/ns/en.sean_penn>
    <http://rdf.freebase.com/ns/type.object.name> "Sean Penn" .
<http://rdf.freebase.com/ns/en.elizabeth_hurley>
    <http://rdf.freebase.com/ns/type.object.name> "Elizabeth Hurley" .
<http://rdf.freebase.com/ns/en.kathryn_bigelow>
    <http://rdf.freebase.com/ns/type.object.name> "Kathryn_Bigelow" .
<http://rdf.freebase.com/ns/en.bob_saget>
    <http://rdf.freebase.com/ns/type.object.name> "Bob Saget" .
<http://rdf.freebase.com/ns/en.ridley_scott>
    <http://rdf.freebase.com/ns/type.object.name> "Ridley Scott" .
<http://rdf.freebase.com/ns/en.russell_crowe>
    <http://rdf.freebase.com/ns/type.object.name> "Russell Crowe" .
<http://rdf.freebase.com/ns/en.mark_strong>
    <http://rdf.freebase.com/ns/type.object.name> "Mark Strong" .
<http://rdf.freebase.com/ns/en.ron_shelton>
    <http://rdf.freebase.com/ns/type.object.name> "Ron Shelton" .
<http://rdf.freebase.com/ns/en.harrison_ford>
    <http://rdf.freebase.com/ns/type.object.name> "Harrison Ford" .
```

```
<http://rdf.freebase.com/ns/en.robert_wagner>
    <http://rdf.freebase.com/ns/type.object.name> "Robert Wagner" .
<http://rdf.freebase.com/ns/en.kurupt>
    <http://rdf.freebase.com/ns/type.object.name> "Kurupt" .
```

SPARQL provides four forms of queries: SELECT, CONSTRUCT, ASK, and DESCRIBE. All of these attempt to find solutions to a graph pattern, and all share similar constructs. While SPARQL is a W3C Recommendation, many semantic platforms support only a few of the SPARQL query forms—however, any semantic platform worth considering should at least support the SELECT form of SPARQL query.

We will describe the uses of and the differences between the four forms, but first let's look at the common structures while we examine the SELECT form.

SELECT Query Form

SPARQL SELECT queries are very similar to the query language we developed in Chapter 3. As with any SPARQL query, a SELECT query can start with a block of PREFIX declarations, which assign shorthand identifiers for URI namespaces that can be used throughout the query. The sample data we will use comes from Freebase, a large, open semantic database that we will cover in the next chapter. To shorten the Freebase URIs, we add the statement:

```
PREFIX fb: <http://rdf.freebase.com/ns/>
```

The initial part of the query can also define a BASE URI to which all relative URIs are concatenated. The following BASE declaration would allow us to use a relative URI of <b006ww0v.rdf> to specify Pete Tong's BBC program, which has an absolute URI of <http://www.bbc.co.uk/programmes/b006ww0v.rdf>:

```
BASE <http://www.bbc.co.uk/programmes/>
```

Note that while you can have any number of PREFIX declarations, you can have at most one BASE declaration.

A SELECT query allows you to identify a subset of the variables used in the graph patterns whose bindings you want returned for each solution. The SELECT clause is followed by a WHERE clause that specifies the graph pattern to match as a collection of triples. Variables in the triple pattern are identified by character strings starting with a question mark (?) or dollar sign ($); there is no difference between the two variable identifiers.

Example 4-1 shows a query that asks which directors have appeared in their own movies.

Example 4-1. Directors who have acted in their own movies

```
PREFIX fb:<http://rdf.freebase.com/ns/>
SELECT ?who ?film
WHERE{
```

```
    ?film fb:film.film.directed_by ?who .
    ?film fb:film.film.starring ?who .
}
```

This query produces the following results:

who	film
fb:en.bob_saget	fb:en.becoming_dick

OPTIONAL and FILTER Constraints

There may be times when you would like to consider information in the solution if it's available, but ignore it if it's not available. This happens frequently when there may be incomplete information about a resource, such as a missing release date in our movie data. SPARQL provides an OPTIONAL clause that allows you to use information in the graph pattern if it's available, but not eliminate solutions if it's missing. For instance, in Example 4-2 we want to list all of Ron Shelton's movies, including the release date if it's available.

Example 4-2. OPTIONAL clause

```
PREFIX fb: <http://rdf.freebase.com/ns/>

SELECT ?film ?reldate
WHERE {
    ?film fb:film.film.directed_by fb:en.ron_shelton .
    OPTIONAL { ?film fb:film.film.initial_release_date ?reldate .}
}
```

When available, the OPTIONAL clause binds the ?reldate to the solution produced by the required part of the graph pattern:

film	reldate
fb:en.dark_blue	
fb:en.hollywood_homicide	2003

Graph patterns are useful for determining whether a specific set of relationships exists within a graph, but frequently you will want to constrain solutions based on specific qualities of a subject, predicate, or object. To do this, SPARQL provides FILTER operations that allow you to specify additional constraints on solution bindings. FILTER constraints use a small set of operators, many derived from XPath 2.0, that allow you to test the variables in the graph pattern for specific conditions. When a FILTER returns false (or an error), the solution under consideration is weeded out.

For instance, while the OPTIONAL clause is useful for handling missing data, you can use OPTIONAL in combination with the **bound** FILTER operator to specifically find subjects that don't assert a specific relationship. Example 4-3 uses FILTER to find films directed by Ron Shelton that do not have a release date.

Example 4-3. Bound FILTER

```
PREFIX fb: <http://rdf.freebase.com/ns/>

SELECT ?film
WHERE {
    ?film fb:film.film.directed_by fb:en.ron_shelton .
    OPTIONAL { ?film fb:film.film.initial_release_date ?reldate .}
    FILTER (!bound(?reldate))
}
```

The result tells us that the dataset does not have a release date for the film *Dark Blue* (fb:en.dark_blue).

SPARQL filter operators also allow you to set up conditions on the qualities of a bound variable's value. For example, it is frequently useful to know whether a string literal follows a specific pattern. For these types of string comparisons, SPARQL provides a regex operator that takes a bound variable, a regex pattern, and an optional set of flags that allow you to do things like ignore case differences between the pattern and the variable's value.

Example 4-4 asks which actors have the string "russell" (case insensitive) in their name.

Example 4-4. REGEX filter

```
PREFIX fb:<http://rdf.freebase.com/ns/>

SELECT distinct ?who ?film
WHERE {
   ?film fb:film.film.starring ?star .
   ?star fb:type.object.name ?who .
      FILTER regex(?who, "russell", "i")
}
```

This yields both Russell Crowe and Kurt Russell. The SPARQL regex operator is defined to be the same as the XPath 2.0 fn:matches operator, which allows us to specify complex regex patterns. For instance, we could limit the results to Russell Crowe by requiring "russell" to be at the beginning of the string using FILTER regex(?who, "^russell", "i").

SPARQL provides a wide variety of comparisons operators for filters, allowing you to partition results. For instance, Example 4-5 uses the inequality filter to find movies released after 2002.

Example 4-5. Inequality filter

```
PREFIX fb:<http://rdf.freebase.com/ns/>

SELECT ?film ?when
WHERE {
```

```
    ?film fb:film.film.initial_release_date ?when .
        FILTER (?when > "2002")
}
```

This returns the following:

film	when
fb:en.hollywood_homicide	2003
fb:en.body_of_lies	2008

Multiple Graph Patterns

Up to this point we have been considering only one graph pattern per query, but SPARQL allows you to specify multiple graph patterns within a query using braces to group triples into separate patterns. Without any modifiers, all patterns are evaluated together to produce the solution (as though all the triple patterns had been in the same group). When pattern groupings are provided, the FILTER clauses apply to the graph pattern they are grouped with.

In Example 4-6, we are looking for directors with three-letter first names and actors whose names start with B. Since both constraints must be satisfied, the solution must be someone who is both an actor and director. We have only one director who is also an actor in our dataset (Bob Saget), so the candidates to our solution set are fairly constrained to start with. Happily, Bob Saget is a director with a three-letter first name and an actor whose name starts with B, so he is the solution to this query.

Example 4-6. Multiple graph patterns

```
PREFIX fb:<http://rdf.freebase.com/ns/>

SELECT ?name
WHERE {

    {
        ?film fb:film.film.directed_by ?person .
        ?person fb:type.object.name ?name
            filter regex(?name, "^... ", "i")
    }

    {
        ?film fb:film.film.starring ?actor .
        ?actor fb:type.object.name ?name
            filter regex(?name, "^b", "i")
    }
}
```

Both graph patterns are evaluated together, producing Bob Saget as the solution.

Using the UNION keyword, we can have each pattern in the query evaluated independently and their solutions joined together. This can be useful when multiple solutions are equally useful. See Example 4-7.

Example 4-7. Multiple graph patterns with UNION

```
PREFIX fb:<http://rdf.freebase.com/ns/>

SELECT ?name
WHERE {

   {
       ?film fb:film.film.directed_by ?director .
       ?director fb:type.object.name ?name
           filter regex(?name, "^... ", "i")
   }
UNION
   {
       ?film fb:film.film.starring ?actor .
       ?actor fb:type.object.name ?name
           filter regex(?name, "^b", "i")
   }
}
```

In this case, there is no requirement that one name fulfill both patterns, so both the directors "Ron Shelton" and "Bob Saget" are solutions.

It is often useful to determine whether two resources are the same when considering a solution. Consider the query in Example 4-8, which looks for directors who have worked with a specific actor, in this case Harrison Ford. The query then asks who else co-starred in those films. It then finds other films in which the director and co-stars worked together. For our results, however, we want to exclude the films in which Harrison Ford worked with the director and co-stars—we are interested only in what other films the director and co-stars have made together. To do this, we specify a FILTER constraint where the ?othermovie and ?movie cannot be the same.

Example 4-8. The usual suspects

```
PREFIX fb: <http://rdf.freebase.com/ns/>

SELECT ?othermovie ?director ?costar
WHERE {
   ?movie fb:film.film.starring fb:en.harrison_ford .
   ?movie fb:film.film.directed_by ?director .
   ?movie fb:film.film.starring ?costar .
   ?othermovie fb:film.film.directed_by ?director .
   ?othermovie fb:film.film.starring ?costar .
      FILTER (?othermovie != ?movie)
}
```

Running this query tells us that Ron Shelton and Kurupt worked with Harrison Ford on a movie, and they also worked together on another movie.

CONSTRUCT Query Form

In many situations it is useful to get a list of variable bindings back from a query, but there are also situations where you want to construct a new graph from the solution set. While you could always write a bit of code that converts solution bindings into tuples and adds them to a graph, SPARQL provides a query form that does this directly. The new graph is constructed from template triples specified in the CONSTRUCT clause, which replaces the SELECT clause. The WHERE and FILTER clauses work in exactly the same way as the SELECT form.

Here we are creating triples indicating who was employed each year based on our movie data:

```
PREFIX fb:<http://rdf.freebase.com/ns/>

CONSTRUCT {
?who <http://employment.history/was_employed_in> ?year
}
WHERE {
    {
       ?film fb:film.film.starring ?who .
       ?film fb:film.film.initial_release_date ?year .
    }
UNION
    {
       ?film fb:film.film.directed_by ?who .
       ?film fb:film.film.initial_release_date ?year .
    }
}
```

ASK and DESCRIBE Query Forms

SPARQL provides a simple ASK form that tests whether a pattern can be found in a graph. The ASK keyword replaces the WHERE keyword, and a simple boolean result is returned indicating whether there is a solution for the pattern in the graph. This query asks whether Bob Saget and Harrison Ford have ever appeared in the same movie:

```
PREFIX fb:<http://rdf.freebase.com/ns/>
PREFIX rdf:<http://www.w3.org/1999/02/22-rdf-syntax-ns#>

ASK {
?film fb:film.film.starring fb:en.bob_saget .
?film fb:film.film.starring fb:en.harrison_ford .
}
```

DESCRIBE is a quirky but potentially powerful query form that, as the SPARQL specification puts it, returns "the useful information the service has about a resource." That is, the results are idiosyncratic to the implementation of the query service. In theory, issuing a DESCRIBE query should help you understand the context of the resources returned, but as they say in consumer disclaimers, "Your results may vary."

In its simplest form, you can ask a query system to describe what it knows about a specific resource:

```
DESCRIBE  <http://rdf.freebase.com/ns/en.harrison_ford>
```

You can use DESCRIBE just as you would SELECT, but instead of getting a set of solution bindings returned, the system will attempt to provide information about the resources returned in the solution. In this example, we want to DESCRIBE the directors who made movies in 2003:

```
PREFIX fb:<http://rdf.freebase.com/ns/>

DESCRIBE ?director
WHERE {

    ?film fb:film.film.initial_release_date "2003" .
    ?film fb:film.film.directed_by ?director .
}
```

While it may seem unsettling to get arbitrary information back about a set of resources, this type of result fits semantic programming patterns extremely well. Unlike traditional programming patterns where the structure (and meaning) of data queries is known when the program is written, in later chapters we will explore an introspective style of programming where programs react to new information as they discover its structure.

SPARQL Queries in RDFLib

The RDFLib Graph class, of which ConjunctiveGraph is a subclass, provides a query method that allows you to run SPARQL queries against your triplestore. The query method takes a string representing the query and an optional initNS keyword parameter that contains a dictionary of namespace mappings. Optionally, you can include the SPARQL prefix declarations directly in your query:

```
from rdflib.Graph import ConjunctiveGraph, Namespace

FBNAMESPACE = Namespace("http://rdf.freebase.com/ns/")
g = ConjunctiveGraph()
g.parse("sample-movie-data.n3", format="n3")

results = g.query("""SELECT ?film ?year
            WHERE { ?film fb:film.film.initial_release_date ?year. }""", \
                initNs={'fb':FBNAMESPACE})

for triple in results:
    print triple
```

You can also run CONSTRUCT queries using RDFLib. The query method returns a resulting graph that can be serialized and used to construct other graphs:

```
from rdflib.Graph import ConjunctiveGraph, Namespace

FBNAMESPACE = Namespace("http://rdf.freebase.com/ns/")
g = ConjunctiveGraph()
```

```
g.parse("sample-movie-data.n3", format="n3")

results = g.query("""CONSTRUCT {
  ?who <http://employment.history/was_employed_in> ?year
}
WHERE {
    ?film fb:film.film.starring ?who .
    ?film fb:film.film.initial_release_date ?year .
  }""", initNs={'fb':FBNAMESPACE}).serialize(format="xml")

print result
```

 If you are having problems with the CONSTRUCT query in RDFLib, check the version number of your build by entering an interactive Python session, importing RDFLib, and entering `rdflib._version_`. If the version is 2.4.0 or earlier, try downloading a more recent version or the trunk of the Subversion (SVN) repository.

To download RDFLib using SVN, enter the following on the command line:

```
$ svn checkout http://rdflib.googlecode.com/svn/trunk/ rdflib-trunk
$ cd rdflib-trunk
$ python setup.py build
```

A fresh build of RDFLib will be available in *build/lib.<yourplatform-architecture>/rdflib*.

You can test the build by changing directories down to the *build/lib.<yourplatform-architecture>* directory and again firing up an interactive Python session. This time when you import RDFLib you should get the new build (which you can check by printing the version of the library).

While RDFLib returns query results in a Python structure, when you run SPARQL queries on other systems you will frequently get your results packed in an XML structure. The W3C has defined a standard XML query result structure that provides a simple set of XML container elements indicating solution sets and the variable bindings within them. See Example 4-9.

Example 4-9. SPARQL XML output

```
PREFIX fb:<http://rdf.freebase.com/ns/>

Select ?film ?year where{
    ?film fb:film.film.initial_release_date ?year .
        FILTER (?year > "2005")
}
```

Here are the results in SPARQL XML output format:

```
<?xml version="1.0"?>
<sparql xmlns="http://www.w3.org/2005/sparql-results#">
```

```
<head>
  <variable name="?film"/>
  <variable name="?year"/>
</head>

<results>
  <result>

    <binding name="year">
        <literal>2008</literal>
    </binding>

    <binding name="film">
        <uri>http://rdf.freebase.com/ns/en.body_of_lies</uri>
    </binding>

  </result>
 </results>
</sparql>
```

SPARQL XML results documents are broken into two parts. The first part, the head, lists each bound variable used in the query within its own variable element. The second part of the document, delineated by a results element, lists the solution sets returned by the query. Individual solutions are surrounded by a result element.

In the example just shown, there is only one solution that binds the variable year to the literal 2008. Optionally, literal elements may also contain datatype or xml:lang attributes. The solution also binds the variable film to the object with the URIref http://rdf.freebase.com/ns/en.body_of_lies.

Useful Query Modifiers

SPARQL is a rich query interface that provides a number of optional modifiers that are very useful when developing real applications. For instance, SPARQL supports a simple form of paginating results using the OFFSET, LIMIT, and ORDER BY solution sequence modifiers. To make pagination work, you must first impose an order on the solutions using the modifier ORDER BY. The ORDER BY modifier takes a list of bound variables, sorts the solutions by the first variable, then sorts the resulting solution sequence further using the second bound variable (if specified), and so on. See Example 4-10.

Example 4-10. Query with two-variable sort

```
PREFIX fb:<http://rdf.freebase.com/ns/>

SELECT ?name ?year
WHERE{
    ?movie fb:film.film.initial_release_date ?year .
    ?movie fb:film.film.starring ?actor .
```

```
        ?actor fb:type.object.name ?name .
} ORDER BY ?year ?name
```

This produces a list of actors sorted by the year they worked on a film, and then by name:

name	year
Bob Saget	2000
Robert Wagner	2000
Elizabeth Hurley	2002
Harrison Ford	2002
Joss Ackland	2002
Sean Penn	2002
Harrison Ford	2003
Kurupt	2003
Robert Wagner	2003
Mark Strong	2008
Russell Crowe	2008

With an order imposed on the solution, you can create pages of specific size using the LIMIT keyword. The OFFSET keyword can then be used to indicate the point in the solution sequence from which to start the next retrieval. If we count pages starting at 1, then in order to obtain the results for page N, specify an OFFSET that is (N-1) * LIMIT. The following code snippet will produce a list of films from our sample dataset, from oldest to newest, paging through the results two solutions at a time:

```
from rdflib.Graph import ConjunctiveGraph

g = ConjunctiveGraph()
g.parse("sparql-example-data.n3", format="n3")

limit = 2
page = 1
results = True

while results:
    print "----page: " + str(page) + "----"

    results = g.query("""PREFIX fb:<http://rdf.freebase.com/ns/>
            SELECT ?film ?year
            WHERE { ?film fb:film.film.initial_release_date ?year. } ORDER BY ?year
                    LIMIT """ + str(limit) + " OFFSET " + str((page-1)*limit))

    for triple in results:
        print triple

    page += 1
```

This taste of SPARQL should give you enough background to formulate useful queries, but it certainly isn't an exhaustive tour of all that is possible with SPARQL. Should you find yourself in need of more information, we suggest taking a look at the W3C Recommendation for SPARQL itself. It's available at *http://www.w3.org/TR/rdf-sparql -query/*, and as specification documents go, it is surprisingly readable with many useful examples.

One final note. Unlike SQL, SPARQL (currently) only supports read operations on the graph, whereas SQL provides update and insert operations. There is no way to modify a graph using SPARQL. Although this is certainly a limitation, it does mean that you can allow untrusted systems (with some limitations on accessible features) access to query infrastructure without fear of your graph being altered. Exposing a raw query interface to a remote data store is a powerful architectural design, and in later chapters we will build applications that not only run SPARQL queries on a local graph store, but also on SPARQL interfaces provided by remote applications.

## CHAPTER 5

Sources of Semantic Data

You now have some tools for storing, querying, and manipulating semantic data. However, none of this is much fun if you don't have any data to put into your triplestore. One of the longest-running criticisms of the semantic web was that no one was publishing data using the standards, so they weren't very useful. Although this certainly held true for a while, these days many more applications, particularly in the social web application realm, are beginning to publish data using semantic web standards.

In this chapter, we will demonstrate how you can obtain and use semantic data from various sources. In doing so, we will also introduce standard vocabularies for describing social networks, music, and movies.

At the end of this chapter, we'll explore Freebase, a semantically enabled social database that provides strong identifiers for millions of entities and vocabularies for hundreds of subject matter domains.

Friend of a Friend (FOAF)

In the previous chapter we introduced FOAF files as an example of how to show the structure of RDF. The FOAF namespace is used to represent information about people, such as their names, birthdays, pictures, blogs, and especially the other people that they know. Thus FOAF files are particularly good for representing data that appears on social networks, and several social networks allow you to access data about their users as FOAF files.

For example, here's a file from hi5, one of the largest social networks worldwide, that is located at *http://api.hi5.com/rest/profile/foaf/358280494*:

```
<rdf:RDF xmlns:hi5="http://api.hi5.com/"
        xmlns:rdf="http://www.w3.org/1999/02/22-rdf-syntax-ns#"
        xmlns:foaf="http://xmlns.com/foaf/0.1/"
        xmlns:rdfs="http://www.w3.org/2000/01/rdf-schema#"
        xmlns:lang="http://purl.org/net/inkel/rdf/schemas/lang/1.1#">
    <foaf:Person rdf:nodeId="me">
      <foaf:nick>Toby</foaf:nick>
```

```
        <foaf:givenName>Toby</foaf:givenName>
        <foaf:surName>Segaran</foaf:surName>
        <foaf:birthday>1-20</foaf:birthday>
        <foaf:img rdf:resource=
            "http://photos3.hi5.com/0057/846/782/gE64Yc846782-01.jpg"/>
        <foaf:weblog rdf:resource="http://blog.kiwitobes.com"/>
        <foaf:gender>male</foaf:gender>
        <lang:masters>en</lang:masters>
        <foaf:homePage rdf:resource=
            "http://www.hi5.com/friend/profile/displayProfile.do?userid=358280494"/>
        <foaf:knows>
          <foaf:Person>
            <foaf:nick>Jamie</foaf:nick>
            <rdfs:seeAlso rdf:resource=
                "http://api.hi5.com/rest/profile/foaf/241087912"/>
          </foaf:Person>
        </foaf:knows>
      </foaf:Person>
   </rdf:RDF>
```

This is Toby's FOAF file from hi5. Since Toby is very unpopular, his only friend is Jamie. The file also provides a lot of other information about Toby, including his gender, birthday, where you can find a picture of him, and the location of his blog. The FOAF namespace, which you can find at *http://xmlns.com/foaf/0.1/*, defines about 50 different things that a file can say about a person.

Many other social networks, such as LiveJournal, also publish FOAF files that can be accessed without signing up for an API key. Because of this, it's almost certain that FOAF files are the most common RDF files available on the Web today.

To reconstruct a portion of the social network from these files, you can build a simple breadth-first crawler for FOAF files. Graph objects from RDFLib have a method called **parse**, which takes a URL and turns it into an RDF graph, so you don't need to worry about the details of the file format. The great thing is that when you parse one FOAF file, you not only get information about one person, but also *the URLs of the FOAF files of all their friends*. This is a very important feature of the semantic web: in the same way that World Wide Web is constructed by linking documents together, the semantic web is made up of connected machine-readable files.

Take a look at the code for a FOAF crawler, which you can download from *http://semprog.com/psw/chapter5/foafcrawler.py*:

```
from rdflib.Graph import Graph
from rdflib import Namespace,BNode

FOAF = Namespace("http://xmlns.com/foaf/0.1/")
RDFS = Namespace("http://www.w3.org/2000/01/rdf-schema#")

def make_foaf_graph(starturi, steps=3):

    # Initialize the graph
    foafgraph = Graph()
```

```
# Keep track of where we've already been
visited = set()

# Keep track of the current crawl queue
current = set([starturi])

# Crawl steps out
for i in range(steps):
    nextstep = set()

    # Visit and parse every URI in the current set, adding it to the graph
    for uri in current:
        visited.add(uri)
        tempgraph = Graph()

        # Construct a request with an ACCEPT header
        # This tells pages you want RDF/XML
        try:
            reqObj = urllib2.Request(uri, None, {"ACCEPT":"application/rdf+xml"})
            urlObj = urllib2.urlopen(reqObj)
            tempgraph.parse(urlObj,format='xml')
            urlObj.close()
        except:
            print "Couldn't parse %s" % uri
            continue

        # Work around for FOAF's anonymous node problem
        # Map blank node IDs to their seeAlso URIs
        nm = dict([(str(s), n) for s, _, n in \
            tempgraph.triples((None, RDFS['seeAlso'], None))])

        # Identify the root node (the one with an image for hi5, or the one
        # called "me")
        imagelist=list(tempgraph.triples((None, FOAF['img'], None)))
        if len(imagelist)>0:
            nm[imagelist[0][0]]=uri
        else:
            nm[''],nm['#me']=uri,uri

        # Now rename the blank nodes as their seeAlso URIs
        for s, p, o in tempgraph:
            if str(s) in nm: s = nm[str(s)]
            if str(o) in nm: o = nm[str(o)]
            foafgraph.add((s, p, o))

        # Now look for the next step
        newfriends = tempgraph.query('SELECT ?burl ' +\
                                'WHERE {?a foaf:knows ?b . \
                                    ?b rdfs:seeAlso ?burl . }',
                                initNs={'foaf':FOAF,'rdfs':RDFS})

        # Get all the people in the graph. If we haven't added them already,
        # add them to the crawl queue
        for friend in newfriends:
            if friend[0] not in current and friend[0] not in visited:
```

```
                    nextstep.add(friend[0])
                    visited.add(friend[0])

            # The new queue becomes the current queue
            current = nextstep
        return foafgraph

    if __name__ == '__main__':

        # Seed the network with Robert Cook, creator of D/Generation
        g = make_foaf_graph('http://api.hi5.com/rest/profile/foaf/241057043', steps=4)

        # Print who knows who in our current graph
        for row in g.query('SELECT ?anick ?bnick '+\
                    'WHERE { ?a foaf:knows ?b . ?a foaf:nick ?anick . ?b \
                        foaf:nick ?bnick . }',
                    initNs={'foaf':FOAF}):
            print "%s knows %s" % row
```

The function make_foaf_graph takes the URI of a FOAF file and the number of steps to search outward as parameters. Don't search too far, or your network will become very large and you may get banned from the service that you're crawling. Notice how we simply give the URI directly to graph.parse, and it takes care of downloading the file and turning it into an RDF graph.

From there, it's easy to query the graph using SPARQL with the namespaces that have been defined (FOAF and RDFS) to find people in the graph and their seeAlso property:

```
SELECT ?burl WHERE {?a foaf:knows ?b . ?b rdfs:seeAlso ?burl . }
initNs={'foaf':FOAF,'rdfs':RDFS}
```

This returns a list of URIs on the right side of the seeAlso property that tell us where to find more information about the people in the graph. If these URIs haven't already been visited, they're added to the queue of URIs that we want to parse and add to the graph.

The main method builds a graph from a starting node and uses a simple query to find all the relationships in the graph and the nicknames (foaf:nick) of the people in those relationships. Try this from the command line:

```
$ python foafcrawler.py
Michael knows Joy
Susan knows Ellen
Michael knows Joe
Mark knows Genevieve
Michael knows James
Michael knows Kimberly
Jon knows John
Michael knows Stuart
Susan knows Jayce
Toby knows Jamie

etc...
```

You can change the starting point and even the social network used by changing the call to `make_foaf_graph`. If you like, you can find out whether your favorite social network supports FOAF and build a graph around yourself and your friends.

Also, remember that although we've used the resources exposed by hi5 in this example, FOAF files can be published by anyone. Rather than joining a social network, you could put a FOAF file on your own web server that connects to other people's files in hi5 or LiveJournal or even FOAF files that they have created themselves. By having a standard way of describing information about people and the relationships between them, it's possible to separate the network of people from the particular site on which it happens to exist.

You can try crawling a distributed social network by starting with Tim Berners-Lee's FOAF page. Change the line that seeds the network to:

```
g=make_foaf_graph('http://www.w3.org/People/Berners-Lee/card',steps=2)
```

Running the code now should crawl out from Tim Berners-Lee, not just within the W3C site, but anywhere his "see also" links point to.

Graph Analysis of a Social Network

Being able to crawl and store graphs such as social networks means you can also apply a little graph theory to understand more about the nature of the graph. In the case of a social network, several questions come to mind:

- Who are the most connected people?
- Who are the most influential people? (We'll see in a moment how "most influential" may differ from "most connected.")
- Where are the cliques?
- How much do people stick within their own social groups?

All of these questions can be explored using well-studied methods from graph theory. Later in this section we'll analyze the FOAF graph, but first we need to get a Python package called NetworkX. You can download NetworkX from *http://networkx.lanl .gov/*, or if you have Python setuptools, install it with `easy_install`:

```
$ easy_install networkx
```

Figure 5-1 shows a simple graph with lettered nodes that we'll use for a first example. In the following Python session, we'll construct that graph and run some analyses on it to demonstrate the different features of NetworkX:

```
>>> import networkx as nx
>>> g = nx.Graph()
>>> g.add_edges_from([('a', 'b'), ('b', 'c'),
... ('b', 'd'), ('b', 'e'), ('e', 'f'), ('f', 'g')])    # Add a few edges
>>> g.add_edge('c','d')                                  # Add a single edge
```

```
>>> nx.degree(g,with_labels=True)                          # Node degree
{'a': 1, 'c': 2, 'b': 4, 'e': 2, 'd': 2, 'g': 1, 'f': 2}

>>> nx.betweenness_centrality(g)                           # Node centrality
{'a': 0.0, 'c': 0.0, 'b': 0.7333, 'e': 0.5333, 'd': 0.0, 'g': 0.0, 'f': 0.3333}

>>> nx.find_cliques(g)                                     # Cliques
[['b', 'c', 'd'], ['b', 'a'], ['b', 'e'], ['g', 'f'], ['f', 'e']]

>>> nx.clustering(g,with_labels=True)                      # Cluster coefficient
{'a': 0.0, 'c': 1.0, 'b': 0.1666, 'e': 0.0, 'd': 1.0, 'g': 0.0, 'f': 0.0}

>>> nx.average_clustering(g)                               # Average clustering
0.30952380952380959
```

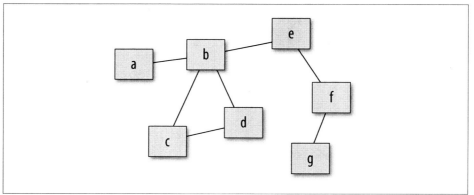

Figure 5-1. A simple graph for analysis

A few different concepts are illustrated in this session. We start by creating a graph and adding the edges to it (a->b, b->c, etc.) so that it represents the graph in Figure 5-1. Then we run a few different analyses on the graph:

degree

 Calculates the *degree* of every node, which is simply the number of nodes connected to this node. It returns a dictionary with every node label and its degree. From the result you can see, for example, that node c has two nodes connected to it.

betweenness_centrality

 Calculates the *centrality* of the node. Centrality is defined as the percentage of shortest paths in the graph that pass through that node—that is, when a message is passed from one random node to another random node, what is the chance that it will have to go through this node? Centrality is sometimes considered a measure of the importance or influence of a node, since it tells how much information must pass through it or how much the network would be disrupted if the node was removed. In this example, node b is the most central. Node e is much more central than node d, even though they both have two neighbors.

find_cliques

Finds all the *cliques* in the graph. A clique is a group of nodes that are all connected to one another, like a tight-knit group of friends in a social network. The smallest cliques have only two members, which just means two nodes are connected. The more interesting cliques are larger—in this case, b, c, and d are all directly connected to one another (b->c, b->d, and c->d), so they form a clique.

clustering

Calculates the *clustering* coefficient of each node. This is a bit more complicated, but it's basically a measure of how cliquish a node is, calculated from the fraction of its neighbors that are connected to one another. In this case, d has a clustering coefficient of 1.0, meaning it is only connected to nodes that are also connected to each other. b, on the other hand, has a coefficient of 0.1666 because even though it's part of the b,c,d clique, it is also connected to other nodes outside the clique.

average_clustering

Just the average of the clustering coefficient of all the nodes in the graph. It's useful as a measure of how cliquish the graph is overall. Social networks tend to be very cliquish, while computer networks are usually not very cliquish at all.

Here's some code for creating a social graph by crawling a set of FOAF files and then running a few NetworkX analyses on it. You can download this file from *http://semprog .com/psw/chapter5/socialanalysis.py*:

```
from rdflib import Namespace
from foafcrawl import make_foaf_graph
import networkx as nx

FOAF = Namespace("http://xmlns.com/foaf/0.1/")
RDFS = Namespace("http://www.w3.org/2000/01/rdf-schema#")

if __name__=='__main__':
    # Build the social network from FOAF files
    rdf_graph = make_foaf_graph('http://api.hi5.com/rest/profile/foaf/241057043', \
        steps=5)

    # Get nicknames by ID
    nicknames = {}
    for id, nick in rdf_graph.query('SELECT ?a ?nick '+\
                                    'WHERE { ?a foaf:nick ?nick . }',
                                    initNs={'foaf':FOAF,'rdfs':RDFS}):
        nicknames[str(id)] = str(nick)

    # Build a NetworkX graph of relationships
    nx_graph = nx.Graph()
    for a, b in rdf_graph.query('SELECT ?a ?b '+\
                                'WHERE { ?a foaf:knows ?b . }',
                                initNs={'foaf':FOAF,'rdfs':RDFS}):
        nx_graph.add_edge(str(a), str(b))
```

```
# Calculate the centrality of every node
cent = nx.betweenness_centrality(nx_graph)

# Rank the most central people (the influencers)
most_connected = sorted([(score, id) for id, score in cent.items()], \
    reverse=True)[0:5]

print 'Most Central'
for score, id in most_connected:
    print nicknames[id], score

print

# Calculate the cluster-coefficient of every node
clust = nx.clustering(nx_graph, with_labels=True)

# Rank the most cliquish people
most_clustered = sorted([(score, id) for id, score in clust.items()], \
    reverse=True)[0:5]
print 'Most Clustered'
for score, id in most_clustered:
    print nicknames[id], score

print
for clique in nx.find_cliques(nx_graph):
    if len(clique) > 2:
        print [nicknames[id] for id in clique]
```

This code builds off the make_foaf_graph function that we defined earlier. After creating the FOAF graph, it creates a table of nicknames (for convenience of displaying results later) and then queries for all the relationships, which it copies to a NetworkX graph object. It then uses some of the NetworkX functions we just discussed to create lists of the most central people (the ones through whom the most information must pass) and the most cliquish people (the ones whose friends all know each other).

Here are some ideas for other things you could try:

- Find the differences between people's rankings according to degree and according to centrality. What are the patterns you noticed that cause people's rankings to differ between these two methods?

- Find and display the cliques in the social network in an interesting way.

- hi5 is a mutual-friendship network; that is, if X is friends with Y, then Y is friends with X. Find an example of a social network where friendships can be one-way and use NetworkX's DiGraph class to represent it.

FOAF was one of the earliest standards, and it's published by a lot of sites, so there's plenty of it on the Web to find, load, and analyze. Public RDF isn't just for social networks, though—in the rest of this chapter you'll see sources for many different types of datasets.

Semantic Search

Crawling around and hoping to find the resource you are looking for, or a predicate relationship you are interested in, may seem a bit haphazard—and it is. Ideally, you are dereferencing a URI because you are interested in the resource.

Like people looking for information on the Web, your application can start its hunt for information by contacting a semantic search engine such as the following:

Yahoo! BOSS

> While you might think of Yahoo!'s Build your Own Search Service (BOSS) as a way of building a better search engine for humans, the BOSS service exposes structured semantic data as well, making it an excellent starting point for semantic applications. As part of its Search Monkey service (described in Chapter 7), Yahoo! extracts RDFa and microformats from every page it indexes. In addition, site owners can provide a DataRSS feed of structured data associated with their sites. The BOSS service makes this structured data available to applications through a simple API that returns XML or JSON structures.
>
> Visit *http://developer.yahoo.com/search/boss/* to learn more about the BOSS service.

Sindice

> Sindice is a mature semantic search engine project sponsored by the Digital Enterprise Research Institute (DERI). Sindice crawls the Web indexing RDF and microformat data and provides a simple API for developer use. Applications can find semantic data sources using a combination of key words, RDF vocabulary constraints, and simple triple patterns. Based on content negotiation, query results are returned in RDF, JSON, or ATOM format.
>
> Learn more about the Sindice service at *http://sindice.com/*.

Freebase

> Unlike the previous systems, which operate more like traditional search engines, Freebase acts as a giant topic hub organized around semantically disambiguated subjects. Freebase can provide you not only with links to other semantic data sources, but also with specific facts about a subject. We will use Freebase in several examples throughout the book, and we will discuss its organization later in this chapter.
>
> See *http://www.freebase.com* for more details.

Linked Data

Crawling from one FOAF document to another, as we did in the previous section, is an example of using "Linked Data." We used the RDF data provided in one graph to guide us to other graphs of data, and as we combined the data from all these graphs, we were able to build a more complete picture of the social network.

This is an example of how strong identifiers can be used to seamlessly join multiple graphs of data. The semantic web, as envisioned by the W3C, is in effect a Giant Global Graph constructed by joining many small graphs of data distributed across the Web. To showcase this type of structure, a community for Linking Open Data (LOD) has emerged, developing best practices around the publication of distributed semantic data.

While RDF provides standard ways for serializing information, the Linking Open Data community has developed standard methods for accessing serialized RDF over the Web. These methods include standard recipes for dereferencing URIs and providing data publishers with suggestions about the preparation and deployment of data on the Web.

While a community of publishers is necessary to bring the Giant Global Graph to fruition, equally important is a community of applications that utilize this collection of distributed semantic data to demonstrate the value of such a graph. This section will explore how to access pieces of the Linked Data cloud and discuss some of the issues involved in building successful Linked Data applications.

As you progress through this book, we hope that you not only find semantic applications easy to build, but also that you see the value of publishing your own data into the Giant Global Graph. And while the Linked Data architecture doesn't currently provide any mechanisms for writing data into the global graph, we will explore Freebase as a semantic data publishing service for Linked Data at the end of this chapter.

The Cloud of Data

As we have seen, FOAF files are a vast distributed source of semantic data about people. There is no central repository of FOAF data; rather, individuals (and systems) that "know" about a specific relationship publish the data in a standardized form and make it publicly available. Some of this data is published to make other data more "findable"—the publisher hoping that by making the data publicly available it will generate more traffic to a specific website as other systems make reference to their information. Others publish information to reduce the effort of coordinating with business partners. And still others publish information to identify sources of proprietary subscription data (in hopes of enticing more systems to subscribe). While there are as many reasons for publicly revealing data as there are types of data, the result is the same: the Internet is positively awash in data.

From a data consumer's perspective, this information is strewn haphazardly about the Web. There is no master data curator; there is no comprehensive index; there is no central coordinator. Harnessing this cloud of data and making it appear as though there were a master coordinator weaving it into a consistent database is the goal of the semantic web.

Smushing Identity

We have made it seem simple to know when two graphs can be merged: either the two graphs will share URIrefs in common or there will be `owl:sameAs` statements identifying how resources in one graph relate to resources in the other graph. In an ideal world, everyone would use the same identifiers for everything, or they would at least know what other identifiers might be used.

But the real world is seldom so tidy. In our ideal world, the publisher of the data would need to know what identifiers other publishers were using. In some cases, when an authoritative source is responsible for cataloging a set of resources (such as biological genes), there is a well-known identifier that everyone can presumably use. But for the vast majority of things we might want to model, this isn't the case.

All is not lost, however; if a publisher can provide even one additional `sameAs` link, it may be possible to find another `sameAs` link in the next graph and so on, from which you can build a collection of alternate identifiers. This approach, while inefficient, does permit an uncoordinated, distributed approach to identity.

But you need not depend on a chain of `sameAs` statements; the context of the data can also provide clues about identity. For instance, many social networks do not provide strong identifiers for members. Instead, they use anonymous nodes when creating FOAF descriptions to represent individuals, assuming that with enough data about the person, you will be able to determine whether you have found the friend you have been searching for.

In many cases this context data is enough to provide a strong hint about the identity of the entity. In FOAF it is often reasonably assumed that the `foaf:mbox` property (or a hash of the `mbox` property) is unique across individuals. When you find two FOAF descriptions that refer to the same `mbox`, you can assume the two descriptions are reporting on the same person.

The effort of bringing identifiers together to claim they refer to the same thing is a well-known problem in several areas of computer science. Because practitioners from many different traditions (and generations) have worked on the problem, people refer to the effort in different ways. Some call it "the identity problem," others call it "reconciliation," while still others call it "record matching." The RDF community, however, has provided the most colorful name for the problem, referring to it as "smushing."

Smushing identity is only part of the challenge—discovering when predicates mean the same thing, or where logical relationships can connect predicates, are other areas of active investigation.

Are You Your FOAF file?

We have been using URIs to identify things in the real world, like people and places—and that's a good thing. But when you request the data from a URI such as `http://semprog.com/ns/people/colin`, you don't really expect to get back a serialized version of Colin. Rather, you expect to get back something like FOAF data that describes Colin.

This subtlety is the reason that many FOAF data generators use anonymous (blank) nodes when describing a person. The FOAF system doesn't have a URI that represents the real person that is distinct from the information resource being produced. This is somewhat like describing something in terms of its attributes without ever naming the object—though obtuse, it does work in many cases. ("I'm thinking of a U.S. President who was impeached and was concerned about what the definition of is was.")

Semantic web architecture makes a distinction between real-world objects, such as Colin, and the information resources that describe those objects, such as Colin's FOAF file. To make this distinction clear, well-designed semantic web systems actually use distinct URIs for each of these items, and when you try to retrieve the real-world object's URI, these systems will refer you to the appropriate information resource.

There are two methods for making the referral from the real-world object to the information resource: a simple HTTP redirect, and a trick leveraging URI fragment (or "hash") identifiers. The redirect method is very general and robust, but it requires configuring your web system to issue the appropriate redirects. The second method is very simple to implement, but it's more limited in its approach.

The HTTP redirect method simply issues an HTTP 303 "see other" result code when a real-world object's URI is referenced. The redirect contains the location of the information resource describing the real-world object. See Figure 5-2.

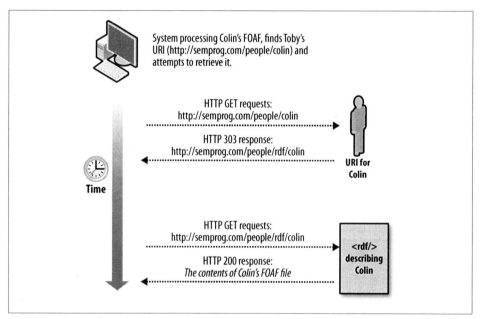

Figure 5-2. Accessing the URI for a real-world object using 303 redirects

The method using fragment identifiers takes advantage of the fact that when an HTTP client requests a URI with a fragment identifier, it first removes the fragment identifier from the URI, thereby constructing a separate URI that it requests from the web system. The URI requested (the URI without the fragment identifier) represents the information resource, which can be delivered by the web system.

We have said that URIs act as strong identifiers, uniquely identifying the things described in RDF statements. By "strong identifier," we mean that you can refer to a resource consistently across *any* RDF statement by using the URI for the resource. No matter where in the universe a URI is used, a specific URI represents one and only one resource. And similarly, a URI represents the same resource over time. A URI today should represent the same resource tomorrow. URIs are not only strong, they should also be stable.

In Chapter 4 we pointed out that every URL is a URI, but how many times have you gone to a URL that used to produce useful information, only to discover that it now produces an HTTP 404 result? If URIs represent strong, stable identifiers, then the information resources produced by dereferencing a URI should also remain stable (or rather, available).

Serving information resources using URIs with fragment identifiers is an easy solution when your RDF is in files. But because URIs should be stable, you must not be tempted to "reorganize" your RDF files should your data expand, as moving resource descriptions between files would change their URIs. It is important to remember that RDF is not document-centric, it is resource-centric, and URIs are identifiers, not addresses. See Figure 5-3.

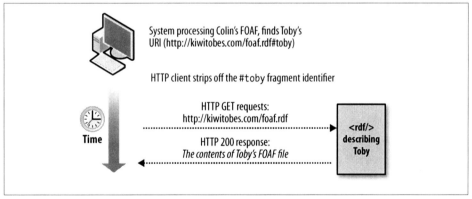

Figure 5-3. Accessing the URI for a real-world object using a fragment identifier

When working with Linked Data, remember that not all URIs can be dereferenced. Although it is unfortunate that we can't learn more about those resources, they still represent useful identifiers. Also, not all URIs representing real-world objects are handled by well-behaved web systems. You will find many examples of RDF information resources being served directly when requesting a real-world object.

Real-World Data

The world is full of surprises, and finding data that doesn't conform to your expectations is always entertaining. In many of our examples we have omitted error checking, or we made assumptions that the data we are working with is well formed. However, when programming with Linked Data, it is important to take a defensive stance and assume that things may not be as advertised.

For instance, when working with RDFa, you may might be surprised to find that not all HTML is well formed (can you believe it?). Some pages may require a tag normalization step (using Tidy or Beautiful Soup) before you can extract the data. Also be prepared for RDF statements that don't use the expected data type. Many sites generate their RDF using database templates, which can lead to systematic errors. So, for example, it is not unusual to find date literals that are marked as `xsd:dateTime` to be written in a human-preferred format. And likewise, you may find literals where you would expect resources, and vice-versa.

So until the cloud of data matures, it is best to "know your data source." Blindly crawling through linked data is an exciting way to program, but you never know what you will get.

So for now, you are not your FOAF file. But perhaps when transporter technology is perfected and humans are assigned a mime type, we will provide an addendum to this book with information on the best practices for retrieving humans through dereferenceable URIs.

Consuming Linked Data

Let's exercise much of what you have learned while walking across a few sources of Linked Data. In this example, we will query multiple data sources, obtaining a critical piece of information from each that will allow us to query the next source. As we query each data source, the information we obtain will be stored in a small internal graph. After we reach the final data source, we will be able to query our internal graph and learn things that none of the sites could tell us on their own.

If this sounds like the "feed-forward inference" pattern we used in Chapter 3, that's no accident. Our rules in this case know how to use identifiers from one data source to obtain data from another data source. Part of what we obtain from each data source is a set of identifiers that can be used with another data source. This type of pattern is very common when working with semantic data, and we will revisit it again and again.

In this section we are going to build a simple Linked Data application that will find musical albums by artists from countries other than the U.S. Our application will take the name of a country as input and will produce a list of artists along with a sample of their discography and reviews of their albums. To find the information required to complete this task, the application will contact three separate data sources, using each to get closer to a more complete answer.

The British Broadcasting Company (BBC) has a wealth of information about music in its archives, including a large collection of record reviews that are not available anywhere else on the Web. Fortunately, the BBC has begun publishing this information as RDF. While the BBC data provides useful information about the significant albums an artist has produced, it provides very little general context data about the artist. In addition, the BBC does not provide a query interface, making it impossible to isolate record reviews of bands that reside in a particular country. But because the BBC's data uses strong identifiers, we can use other Linked Data to find the information we want.

The BBC data uses identifiers provided by the MusicBrainz music metadata project. The MusicBrainz project is a well-regarded community effort to collect information about musical artists, the bands they play in, the albums they produce, and the tracks on each album. Because MusicBrainz is both a well-curated data collection and a technology-savvy community, its identifiers are used within many datasets containing information about musical performances.

MusicBrainz itself does not provide Linked Data dereferenceable URIs, but Freebase —a community-driven, semantic database that we will look at later—uses MusicBrainz identifiers and provides dereferenceable URIs. Freebase also connects the MusicBrainz identifiers to a number of other strong identifiers used by other data collections.

DBpedia, an early Linked Data repository, is an RDF-enabled copy of Wikipedia. Freebase and DBpedia are linked by virtue of the fact that both systems include Wikipedia identifiers (so they are able to generate `owl:sameAs` links to one another). DBpedia also provides a SPARQL interface, which allows us to ask questions about which Wikipedia articles discuss bands that reside in a specific country. From the results of this query, we will follow the Linked Data from one system to the next until we get to the BBC, where we will attempt to find record reviews for each of the bands. See Figure 5-4.

You can build this application as we work through the code. Alternatively, you can download it, along with other demonstrations of consuming Linked Data, from *http://semprog.com/psw/chapter5/lod*.

Let's start by defining the namespaces that we'll be using throughout the application and constructing a dictionary of the namespace prefixes to use across our queries. We will also create a few simple utility functions for submitting requests via HTTP to external services.

 Caveat dereferencer! This Linked Data application depends on the availability of three independent web services. Over time the functionality, or even existence, of these services may change. If you run into problems with this or any other Linked Data application, try sending a few requests manually to each service using wget or curl to see if the service is working as you expect.

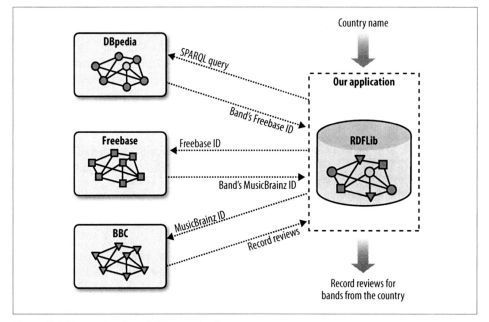

Figure 5-4. Traversing three LOD repositories

```
"""
Example of using Linked Data
1) from DBpedia: get bands from a specific country
2) from Freebase: get Musicbrainz identifiers for those bands
3) from the BBC: get album reviews for those bands
"""

import urllib2
from urllib import quote
from StringIO import StringIO
from rdflib import Namespace, Graph, URIRef

countryname =  "Ireland" #also try it with "Australia"
dbpedia_sparql_endpoint =
    "http://dbpedia.org/sparql?default-graph-uri=http%3A//dbpedia.org&query="

#namespaces we will use
owl = Namespace("http://www.w3.org/2002/07/owl#")
fb = Namespace("http://rdf.freebase.com/ns/")
foaf = Namespace("http://xmlns.com/foaf/0.1/")
rev = Namespace("http://purl.org/stuff/rev#")
dc = Namespace("http://purl.org/dc/elements/1.1/")
rdfs = Namespace("http://www.w3.org/2000/01/rdf-schema#")

nsdict = {'owl':owl, 'fb':fb, 'foaf':foaf, 'rev':rev, 'dc':dc, 'rdfs':rdfs}

#utilities to fetch URLs
def _geturl(url):
```

```
        try:
            reqObj = urllib2.Request(url)
            urlObj = urllib2.urlopen(reqObj)
            response = urlObj.read()
            urlObj.close()
        except:
            #for now: ignore exceptions, 404s, etc.
            print "NO DATA"
            response = ""
        return response

    def sparql(url, query):
        return _geturl(url + quote(query))

    def fetchRDF(url, g):
        try: g = g.parse(url)
        except: print "fetch exception"
```

Next we will start defining functions that operate on specific data sources. Each function will take a reference to our internal graph as an argument, then determine if there is something new it can add to the data by contacting an external source. If so, it contacts the external source and adds the data to the internal triplestore. If our functions are well-written, they should only contact the external source when they detect that there is information missing in the local graph and they know that the external source will supply it. Much like the multi-agent blackboard described in Chapter 3, we should be able to call the functions repeatedly and in any order, allowing them to opportunistically fill in information when they are called.

To start the process, we will define a function that queries DBpedia with our one input parameter, the name of a country. DBpedia will provide a list of rock bands originating from that country and supply us with their Freebase identifiers:

```
def getBands4Location(g):
    """query DBpedia to get a list of rock bands from a location"""

    dbpedia_query = """
        PREFIX rdf: <http://www.w3.org/1999/02/22-rdf-syntax-ns#>
        PREFIX owl: <http://www.w3.org/2002/07/owl#>
        PREFIX dbpp: <http://dbpedia.org/property/>
        PREFIX dbpo:<http://dbpedia.org/ontology/>
        PREFIX dbpr:<http://dbpedia.org/resource/>

        CONSTRUCT {
          ?band owl:sameAs ?link .
        } WHERE {
          ?loc dbpp:commonName '""" + countryname + """'@en .
          ?band dbpo:homeTown ?loc .
          ?band rdf:type dbpo:MusicalArtist .
          ?band dbpo:genre dbpr:Rock_music .
          ?band owl:sameAs ?link .
            FILTER regex(?link, "freebase")
        }"""
```

```
print "Fetching DBpedia SPARQL results (this may take a few seconds)"
dbpedia_data = sparql(dbpedia_sparql_endpoint, dbpedia_query)
g.parse(StringIO(dbpedia_data),format='xml') #put results in local triplestore
print "done with dbpedia query"
```

Next we will define a function that looks for bands that have a Freebase identifier but do not have information about the band's name filled in. For each band in this state, the function will contact Freebase and add what it learns to our internal graph:

```
def getMBZIDs(g):
    """Query the local triplestore to find the Freebase links for each band
    and load the Freebase data for the band into the local triplestore"""

    #Freebase provides the canonical name for each band,
    #so if the name is missing, we know that we haven't asked Freebase about it
    fbquery = """
        SELECT ?fblink WHERE{
            ?band owl:sameAs ?fblink .
            OPTIONAL { ?fblink fb:type.object.name ?name . }
            FILTER regex(?fblink, "freebase", "i")
            FILTER (!bound(?name))
    }"""

    freebaserefs = [fref[0] for fref in g.query(fbquery, initNs=nsdict)]

    print "Fetching " + str(len(freebaserefs)) + " items from Freebase"
    for fbref in freebaserefs:
        fetchRDF(str(fbref), g)
```

Our next functions will look for bands that have a BBC identifier but for which we have no review information. The first function retrieves artist information from the BBC to obtain information about the albums the artist has made. The second function looks for albums that don't have review information and retrieves the text of the review from the BBC archive:

```
def getBBCArtistData(g):
    """For each MusicBrainz ID in the local graph try to retrieve a review from
    the BBC"""

    #BBC will provide the album review data,
    #so if its missing we haven't retrieved BBC Artist data
    bbcartist_query = """
SELECT ?bbcuri
WHERE {
    ?band owl:sameAs ?bbcuri .
    OPTIONAL{
        ?band foaf:made ?a .
        ?a dc:title ?album .
        ?a rev:hasReview ?reviewuri .
    }
    FILTER regex(?bbcuri, "bbc", "i")
    FILTER (!bound(?album))
}"""

    result = g.query(bbcartist_query, initNs=nsdict)
```

```
        print "Fetching " + str(len(result)) + " artists from the BBC"

        for bbcartist in result:
            fetchRDF(str(bbcartist[0]), g)

def getBBCReviewData(g):

    #BBC review provides the review text
    #if its missing we haven't retrieved the BBC review
    album_query = """
SELECT ?artist ?title ?rev
WHERE {
    ?artist foaf:made ?a .
    ?a dc:title ?title .
    ?a rev:hasReview ?rev .
}"""
    bbc_album_results = g.query(album_query, initNs=nsdict)
    print "Fetching " + str(len(bbc_album_results)) + " reviews from the BBC"

    #get the BBC review of the album
    for result in bbc_album_results:
        fetchRDF(result[2], g)
```

Finally, we will sequence the use of our "rules" in the body of our application. We start by populating a local graph with the results of our DBpedia SPARQL query that identifies bands from the country of our choosing. Next we call each of our rule functions to fill in missing data.

While we call the functions in an obvious sequence, you could rework this section to loop over the calls to each data acquisition function, emitting completed review data as it is obtained. In theory, should a data source become unavailable, the application would just keep iterating through the rules until the data became available, allowing the review to be retrieved. Similarly, you should be able to modify the body of this application so that you can "inject" new countries into the application and have the application kick out additional reviews as new input becomes available:

```
if __name__ == "__main__":

    g = Graph()
    print "Number of Statements in Graph: " + str(len(g))

    getBands4Location(g)
    print "Number of Statements in Graph: " + str(len(g))

    getMBZIDs(g)
    print "Number of Statements in Graph: " + str(len(g))

    getBBCArtistData(g)
    print "Number of Statements in Graph: " + str(len(g))

    getBBCReviewData(g)
    print "Number of Statements in Graph: " + str(len(g))
```

```
final_query = """
SELECT ?name ?album ?reviewtext
WHERE {
    ?fbband fb:type.object.name ?name .
    ?fbband owl:sameAs ?bband .
    ?bband foaf:made ?bn0 .
    ?bn0 dc:title ?album .
    ?bn0 rev:hasReview ?rev .
    ?rev rev:text ?reviewtext .
    FILTER ( lang(?name) = "en" )
}"""

finalresult = g.query(final_query, initNs=nsdict)
for res in finalresult:
    print "ARTIST: " + res[0] + " ALBUM: " + res[1]
    print "----------------------------"
    print res[2]
    print "============================="
```

This application makes use of Freebase as a semantic switchboard, exchanging one identifier for another. Freebase can also provide a map of data resources that use specific types of identifiers, and it can itself be a source of information about a wide variety of subjects. In the next section we will look at Freebase in greater depth and examine some of the services it provides to make writing semantic applications a bit easier.

Goal-Driven Inference

While we talk about feed-forward inferencing in this example and others throughout the book, you should also be aware of another version of this application that starts with the goal of producing record reviews from the BBC and operates on a map of linked data sources to find a path of requests that will accomplish our goal.

This type of inference is known as *backward chaining reasoning*—you start with a goal, and then search through the inference rules to see which rules produce outcomes consistent with that goal. Once you have found a set of rules that can produce the goal state, the new goal becomes finding the rules that would allow the first set of rules to fire. You keep looking for outcomes that fulfill the current goal and trying to set up the preconditions for those rules, and so on until you reach the current state of the system (or in our case, the input state).

We will use feed-forward inferencing here, as it is well suited to situations where conditions are constantly changing, such as systems going offline or new data being added by users.

Freebase

Freebase is an open, writable, semantic database with information on millions of topics ranging from genes to jeans. Within Freebase, you can find data from international and government agencies, private foundations, university research groups, open source data projects, private companies, and individual users—in short, anyone who has made their data freely available. And as the name suggests, Freebase doesn't cost anything to use. All the data in it is under a Creative Commons Attribution (CC-BY) license, meaning it can be used for any purpose as long as you give credit to Freebase.

Like the structures we have been examining, Freebase is a graph of data, made up of nodes and links. But unlike the RDF graphs we have looked at so far, Freebase envelops this raw graph structure with a query system that allows developers to treat these structures as simple objects that are associated with one or more Freebase types. A type, in Freebase, is a collection of properties that may be applied to the object, linking it to other objects or literal values.

An Identity Database

Freebase has information on millions of entities (or "topics"), and the data is actively curated by the community of users (and a wide variety of algorithmic bots) to ensure that each entity is unique. That is, each Freebase topic represents one, and only one, semantically distinct "thing." Said another way, everything within Freebase has been reconciled (or smushed!). In theory, you should never find duplicated topics, and when they do occur, the community merges the data contained in the two topics back into one. This active curation is a part of what makes Freebase useful as an identity database. When you dereference a Freebase identifier, you will get back one and only one object. All information for the topic, including all other identifiers, are immediately available from that one call.

As a semantic database, strong identifiers play a central role within Freebase. Every topic has any number of strong, stable identifiers that can be used to address it. These identifiers are bestowed on topics by Freebase users and by external, authoritative sources alike. Topic identifiers are created by establishing links from a topic to special objects in Freebase called namespaces. The links between topics and namespaces also contain a key value. The ID for a topic is computed by concatenating the ID of a parent namespace, a slash (/), and the value of the key link connecting the topic to the namespace. For instance, the band U2 is represented by the topic with the key value of `a3cb23fc-acd3-4ce0-8f36-1e5aa6a18432` in the namespace with the ID `/authority/musicbrainz`. The ID for the U2 topic can thus be represented as `/authority/musicbrainz/a3cb23fc-acd3-4ce0-8f36-1e5aa6a18432`. The MusicBrainz namespace ID can similarly be understood as having a key with the value `musicbrainz` in the namespace with the ID `/authority`, and so on back to the root namespace.

As we will see, Freebase IDs are useful within the Freebase system and also serve as externally dereferenceable URIs. This means that not only can you refer to any Freebase topic in your own RDF, but any data you add to Freebase can be referenced by others through RDF.

GUIDs or IDs?

Freebase is actually an append-only graph store. On the surface, data in Freebase appears to be mutable, but that is a convenient illusion created by the default read API. In truth, once a piece of data is written into Freebase, it is (in theory) forever findable. This immutability has an interesting implication: the state of the graph at any point in the past can be read. This allows you to see all changes to a topic by inspecting the history of all links connecting to that node in the graph. It also means you can ask for query results based on the state of the graph at a specific point in time.

In addition to any number of keys that can be used as an ID for a given topic, Freebase also provides a 32-character GUID. While this GUID can be used as an ID, it doesn't refer to the Freebase topic, but rather it represents the underlying graph object holding the links for the topic at this point in time. The difference between a key-based ID and a GUID is subtle but important. If a topic is merged (or split), the keys on the topic move with the semantics—they track the meaning of the topic. But the GUID remains fixed to the graph object. When two topics are merged, one of the GUIDs will become empty. Unless you are interested in tracing the history of a topic, you will almost always want to use key-based IDs to reference topics.

RDF Interface

As we saw earlier in the Linked Data section, Freebase provides an RDF Linked Data interface, making Freebase a part of the Giant Global Graph. As a community-writable database, this means that any data (or data model) within Freebase is immediately available as Linked Data, and thus you can use Freebase to publish semantic data for use in your own Linked Data applications.

You can construct a dereferenceable URI representing any object in Freebase by taking an ID for the object, removing the first slash (/), replacing the subsequent slashes with dots (.), and appending this transformed ID on the base URI http://rdf.freebase.com/ns/. For instance, the actor Harrison Ford (the one who appeared in *Star Wars*) has the Freebase ID /en/harrison_ford. You can find out what Freebase knows about him by dereferencing the URI http://rdf.freebase.com/ns/en.harrison_ford.

These URIs represent the things Freebase has information about, so the URI http://rdf.freebase.com/ns/en.harrison_ford represents Harrison Ford the actor. Since Freebase can't produce Harrison Ford the actor, but it can provide information about him, this URI will be redirected with an HTTP 303 response to the URI for the information resource about him. If your request includes an HTTP ACCEPT header indicating a preference for application/rdf+xml content, you will be redirected to the information resource at http://rdf.freebase.com/rdf/en/harrison_ford. (You can also request N-triples, N3, and turtle with the appropriate ACCEPT headers.) If you make the request with an ACCEPT header (or preference) for text/html, you will be redirected to the standard HTML view within Freebase.com.

Freebase Schema

Not only does Freebase allow users to add data to the graph, it also allows them to extend the data model to fit their data. Data models within Freebase are called *schemas* and are broken down along areas of interest called *domains*. Domains serve only to collect components of a data model; they have no specific semantic value.

In Freebase, any object can have one or more *types*. Types provide a basic categorization of objects within Freebase and indicate what *properties* you might expect to find linked to the object. Since Freebase properties specify the link between two objects, they serve the same role as RDF predicates (in fact, the Freebase RDF interface uses properties as predicates). Unlike an object-oriented programming language, or some of the RDF models we will examine in later chapters, Freebase types do not have inheritance. If you want to say that something is also a more abstract type, you must explicitly add the type to the object.

For instance, in Freebase, Harrison Ford is of type `actor`, which resides in the `film` domain (/film/actor), and he is also typed as a `person`, which resides in the `people` domain (/people/person). The `person` type has properties such as `date_of_birth`, `nationality`, and `gender`. As in RDF, these properties represent links connected to the object representing Harrison Ford. Properties such as `date_of_birth` represent literals, whereas properties for things like `nationality` and `gender` represent links to other objects. Properties in Freebase not only indicate whether they are links to literals or other objects, but they also specify the type of object at the other end of the link. Therefore, you know that when you reference the `gender` property on the `person` schema, you will find an object with the `gender` type on the other end of the link. See Figure 5-5.

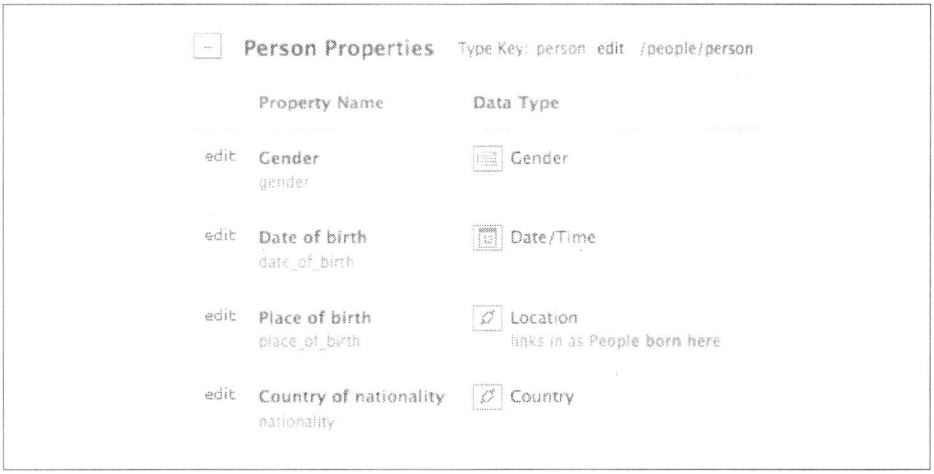

Figure 5-5. Freebase /people/person schema (http://www.freebase.com/type/schema/people/person)

Similarly, the `actor` type has a `film` property. It too is a link to another object, but rather than linking directly to a `film` object, it connects to a `performance` object that is itself connected to a film. This `performance` object serves the same purpose as a blank node in RDF, allowing us to express information about the relationship between two entities—in this case, allowing us to identify the character the actor played in the film. Unlike an RDF model, however, these "mediating" nodes are first-class objects in Freebase and therefore have strong, externally referenceable identifiers. See Figure 5-6.

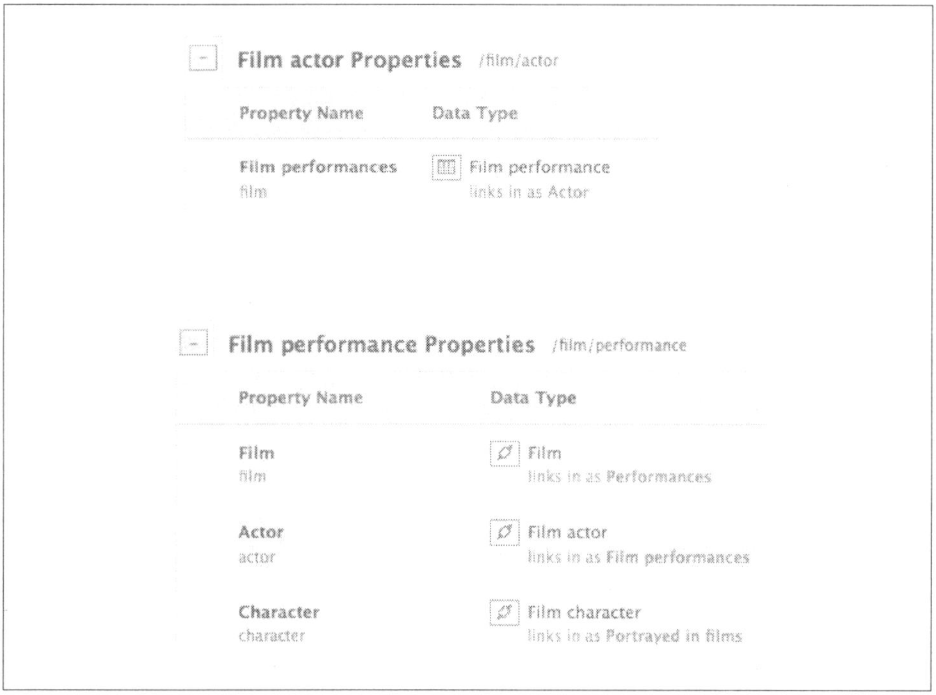

Figure 5-6. Freebase /film/actor and /film/performance schemas

As you might have guessed from our examples, the ID of a type in Freebase is the domain that the type is contained in, followed by a key for that type. That is, a domain (such as `people`) is also a namespace, and types have keys (e.g., `person`) in domains (giving the `people` type the ID `/people/person`). Similarly, types operate as namespaces for properties, so the full ID for the birthday property is `/people/person/date_of_birth`.

Every object in Freebase is automatically given the type `object` (`/type/object`), which provides properties available to all objects, such as `name`, `id`, and `type`. The `type` property (`/type/object/type`) is used to link the object with type definitions. For instance, Harrison Ford is an object of type `actor`, by virtue of having a `/type/object/type` link to the `/film/actor` object. In a self-similar fashion, types themselves are nothing more than objects that have a `/type/object/type` link to the root type object `/type/type`. See Figure 5-7.

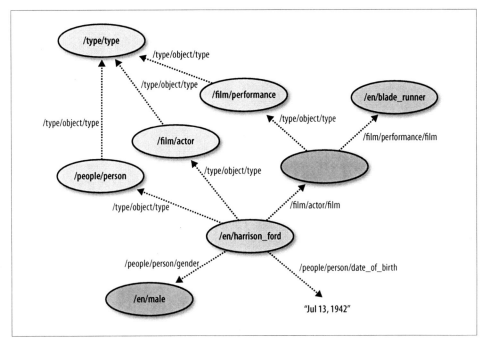

Figure 5-7. Links from the /en/harrison_ford object in Freebase

Since the RDF interface can provide information about any object in Freebase, you can ask for an RDF description of each type used by the /en/harrison_ford object. The type description will include information about the properties used by the type, and you can continue the investigation by asking the interface for the definition of specific properties in RDF.

For example, we can ask about the /film/actor type using the wget command-line tool:

```
wget -q -O - --header="ACCEPT:text/plain" http://rdf.freebase.com/ns/film.actor
```

This returns a triple that tells us that /film/actor has a property called film (/film/actor/film):

```
<http://rdf.freebase.com/ns/film.actor>
        <http://rdf.freebase.com/ns/type.type.properties>
                <http://rdf.freebase.com/ns/film.actor.film>.
```

We can then ask for the definition of the film property with:

```
wget -q -O - --header="ACCEPT:text/plain" http://rdf.freebase.com/ns/film.actor.film
```

MQL Interface

Freebase uses a unique query language called MQL (pronounced like nickel, but with an M). Unlike SPARQL's graph pattern approach, MQL uses tree-shaped query-by-example structures to express queries. With MQL, you indicate the node and link arrangement you are searching for, filling in constraints along the structure and leaving blank slots where you want results returned. The query engine then searches the graph and returns a list of all structures that match the shape of the query and the embedded constraints. The query engine will also complete the structure, filling in any blank slots with information from the graph for each match it finds.

MQL query structures are expressed using JavaScript Object Notation (JSON), making them very easy to construct and parse in Python. JSON array structures are used where a result component may have multiple values, and JSON objects are used to select named properties. For instance, to discover Harrison Ford's birthday, you would write:

```
{"id":"/en/harrison_ford, "/people/person/date_of_birth":null}
```

MQL allows you to use the property key to shorten the expression if the type of the object is explicitly declared as a constraint in the query. Thus, the query just shown can also be expressed as:

```
{"id":"/en/harrison_ford", "type":"/people/person", "date_of_birth":null}
```

F8: The (Hidden) Developer Bar

While exploring the Freebase system, you can press F8 to toggle a small blue ribbon at the bottom of every page. This "developer toolbar" provides a number of useful links for exploring the data (and data model) behind the standard Freebase views. For instance, while looking at a Freebase topic, the "Explore" link will take you to a page that shows a "tuple" view of the graph data for the topic being displayed, along with additional information that isn't normally displayed in the standard UI (such as the list of keys and namespaces for the topic).

The developer toolbar also provides a link to the Freebase Query Editor (*http://www .freebase.com/tools/queryeditor*), where you can experiment with these MQL queries yourself.

Similarly, since the type of object at the other end of a property link is known via the schema, when specifying properties of objects returned by properties "higher up" in the query tree, you can use their shortened IDs. For instance, if we ask for the `film` property on Harrison Ford, when treated as a `/film/actor`, we would retrieve all his `/film/performance` objects. Since these `performance` objects themselves aren't that interesting, we will want to obtain the `film` property for each of these `performance` objects. Because the query processor knows the type of each of these objects, we can ask for them using just the property key:

```
{
  "id":"/en/harrison_ford",
  "type":"/film/actor",
  "film":[{ "film":[] }]
}
```

When a constraint is left empty in a query, in most cases Freebase will fill the slot with the name of the object or the value of the literal that fills the slot. One exception to this is when an empty constraint would be filled by an object used by the type system; in these cases, the ID of the object is returned.

We can, however, tell Freebase exactly what we want to see from each object returned by using curly braces ({}) to expand the object and call out specific properties in the dictionary. Thus, to get the names of the movies Harrison Ford has acted in and the directors of those movies, we can write:

```
{
  "id":"/en/harrison_ford",
  "type":"/film/actor",
  "film":[{ "film":[{"name":null, "directed_by":[] }] }]
}
```

And we can expand the director object further, constraining our query to find only female directors who have worked with Harrison Ford (note that we must use the full property ID for the gender constraint since the directed_by property returns a /film/director object):

```
[{
  "id":"/en/harrison_ford",
  "type":"/film/actor",
  "film":[{ "film":[{"name":null,
      "directed_by":[{"name":null, "/people/person/gender":"Female"}] }] }]
}]
```

MQL has many additional operators that allow you to further constrain queries, but this quick introduction should give you enough background on the Freebase data model to start making queries on your own.

Using the metaweb.py Library

Calls are made to the Freebase services using HTTP GET requests (or optionally POST for some services when the payload is large). But rather than learning the various parameters for each call, it is easier to grab a copy of the *metaweb.py* library that handles the HTTP requests for you.

The *metaweb.py* library covered in this section is available from *http://www.freebase .com/view/en/appendix_b_metaweb_py_module* and makes use of the *simplejson* library for encoding and decoding JSON structures. You can download the *simplejson* library from *http://code.google.com/p/simplejson*.

 Enhanced versions of *metaweb.py* are posted on Google Code; they provide additional features and access to lower-level APIs should you need them in the future.

Place a copy of *metaweb.py* in your working directory and start an interactive Python session. Let's start by making the Harrison Ford query that lists all his movies:

```
>> import metaweb
>> null = None
>> freebase = metaweb.Session("api.freebase.com") #initialize the connection
>> q = {"id":"/en/harrison_ford", "type":"/film/actor", "film":[{ "film":null }] }
>> output = freebase.read(q)
>> print str(output)

{u'type': u'/film/actor', u'id': u'/en/harrison_ford', u'film': [{u'film':
    [u'Air Force One']},
{u'film': [u'Apocalypse Now']}, {u'film': [u'Blade Runner']}, {u'film':
    [u'Clear and Present Danger']},
{u'film': [u'Firewall']}, {u'film': [u'Frantic']}, {u'film':
    [u'Hollywood Homicide']},...
```

Of course, we can iterate over pieces of the result structure to make the output a bit more clear:

```
>> for performance in output['film']: print performance['film']

Air Force One
Apocalypse Now
Blade Runner
Clear and Present Danger
Firewall
Frantic
Hollywood Homicide
....
```

By default, Freebase will return the first 100 results of a query, but MQL provides a cursor mechanism that lets us loop through pages of results. The *metaweb.py* library provides a `results` method on the `session` object, which returns a Python iterator that makes cursor operations transparent.

This time, let's look at all films in Freebase that list a female director (notice that the query is surrounded by square brackets, indicating that we believe we should get back multiple results):

```
>> q2 = [{"name":null, "type":"/film/film",
    "directed_by":[{ "/people/person/gender":"Female" }] }]
>> gen = freebase.results(q2)
>> for r in gen:
...    print r["name"]

15, Park Avenue
3 Chains o' Gold
30 Second Bunny Theater
36 Chowringee Lane
A League of Their Own
A New Leaf
A Perfect Day
Une vraie jeune fille
....
```

In addition to packaging access to the query interface, *metaweb.py* provides a number of other methods for accessing other Freebase services. For instance, if you query for properties that produce objects of type /common/content, you can fetch the raw content (text, images) using the blurb and thumbnail methods. Try looking at the documentation in the source for the search method and see whether you can find topics about your favorite subject.

Interacting with Humans

Unlike machines, humans do well with ambiguous names, which can make it challenging when soliciting input from humans for semantic applications. If you provide a simple input box asking United States residents to enter the country of their residence, you will obtain input ranging from simple acronyms like "US" to phrases like "The United States of America", all of which ostensibly identify the country <http://rdf.freebase.com/ns/en.united_states>. The problem is compounded when users enter semantically ambiguous data. For example, it's likely that users who "idolize" Madonna form more than one demographic.

One obvious solution is to create an enumeration of "acceptable" responses that are equated with strong identifiers. Although this will work, unless you generate an extremely large list of options, you will substantially limit your users' expressive power. With over 14.1 million unique, named entities, Freebase provides a sufficiently diverse yet controlled lexicon of things users might want to talk about—and each comes with a strong identifier.

To facilitate the use of Freebase for human input, the "autocomplete" widget used within the Freebase site has been made available as an open source project. With about 10 lines of JavaScript, you can solicit strong identifiers in a user-friendly way. To see how easy it is, let's build a very simple web page that will generate the RDF URI for a Freebase topic. Open your favorite plain-text editor and create the following page:

```
<head>

<script type="text/javascript"
    src="http://ajax.googleapis.com/ajax/libs/jquery/1.3/jquery.min.js">
</script>

<script type="text/javascript"
    src="http://controls.freebaseapps.com/suggest"></script>
<link rel="stylesheet" type="text/css"
    href="http://controls.freebaseapps.com/css" />

<script>
  function getRDF(){
    var freebaseid = document.getElementById("freebaseid").value;
    var rdfid = freebaseid.substr(1).replace(/\//g,'.');
    var rdfuri = "http://rdf.freebase.com/ns/" + rdfid;
    document.getElementById("rdfuri").innerHTML = rdfuri;
  }
```

```
    </script>
  </head>

  <body>
    <form>
        URIref for: <input type="text" id="topicselect" />
        <input type="hidden" name="freebaseid" id="freebaseid">
        <input type="button" onclick='getRDF()' value="Fetch Data">
    </form><p>

    RDF URI: <span id="rdfuri"></span><p>

    <script type="text/javascript">
        $(document).ready(function() {
            $('#topicselect').freebaseSuggest({})
            .bind("fb-select", function(e, data)
                {$('#freebaseid').val(data.id); getRDF();});
        });
    </script>
  </body>
```

Open the page in your web browser and start typing the word "Minnesota" into the text box. Freebase Suggest will bring up a list of topics related to the stem of the word as you are typing. Once you have the selected the topic of interest, jQuery fires the function bound to `fb-select`. In this case, the Freebase ID (`data.id`) of the topic selected is placed in the hidden text input box (`$('#freebaseid')`). When you click the "Fetch Data" button, a simple JavaScript function reformats the Freebase ID into the Freebase RDF ID and provides a link to the Freebase Linked Data interface.

The Freebase Suggest code provides a number of simple hooks that allow you to customize not only the appearance of the widget, but also how it behaves. For instance, you can provide custom filters to limit the topics presented and custom transformations to display different pieces of information specific to your task.

We have barely scratched the surface of Freebase in this book; Freebase provides various other services to facilitate application development, and it provides a complete JavaScript development environment for building customized APIs and small applications. You can learn more about these and other Freebase services at *http://www .freebase.com*.

Throughout this chapter we've introduced new vocabularies, such as FOAF and Freebase Film. We've provided a page of vocabularies at *http://semprog.com/docs/vocabula ries.html*, which we'll keep updated with the ones we think are useful and interesting. In the next chapter, we'll look at how to go about creating an *ontology*, which is a formal way of defining a semantic data model.

## CHAPTER 6

What Do You Mean, "Ontology"?

We stated in Chapter 1 that the role of semantics was to communicate enough meaning to result in an action. Over the last few chapters we have shown how to represent and transmit knowledge in a formal, machine-readable way. Throughout our exploration of triples and graphs, we have relied on existing sets of predicates or simply invented new ones, using our intuitions and shared experiences about subject and object relations to guide us.

This informal approach to modeling our data has served us well, and is frequently sufficient for small, independent projects. We can write programs that respond to predicates like `foaf:knows` or `fb:film.film.starring` in reliable ways, and from the actions of those programs we might state that our program "knows what it means to star in a film." But our definition of "film" is murky at best. Should our set of films include made-for-TV movies? Does a movie being included in our set of films mean it was eligible for an Academy Award? These questions may seem pedantic, but as projects grow and information is distributed more widely, having a precise way to represent and transmit this understanding becomes increasingly important.

In this chapter we examine how to build more complete models of relationships and how to express the models in RDF itself. In later chapters, when we are building larger and more sophisticated programs, we will leverage these formal models not only to define data relationships, but also to dynamically drive applications.

What Is It Good For?

The last few chapters have demonstrated that expressions using subjects, predicates, and objects are a useful and flexible method for representing knowledge. The discussion of entities and relationships—the subjects and objects we can talk about, and the predicates describing how entities relate to one another—has traditionally been within the purview of philosophy. Philosophers since Aristotle have been greatly concerned about knowing what exists and how to describe it. And once you think you know that something exists, the next step is to find a place for it among all other things. In the domain of philosophy, this effort is called *ontology*.

But why should philosophers have all the fun? Shouldn't computer programmers be a part of this franchise, too?

To the degree that semantic software systems need a model of the world from which to make sense of the knowledge that they operate on, we as system designers are also ontologists. But while the philosophers are trying to sort out the order of the universe, we need only concern ourselves with creating order and describing relationships of things important to our applications.

Building models of the world is nothing new to system designers—we do this all the time, whether we are creating database schemas or specifying object relations. But with semantic systems, we need to express these models in such a way that systems distributed across the Web can read our models and understand with precision how to use them.

A Contract for Meaning

Does it matter if our list of films includes made-for-TV movies? Probably not. Though the film *Sarah T.—Portrait of a Teenage Alcoholic* may not have had a theatrical release, the film was produced like any theatrical movie of its time, and the people who worked on it were well-known actors and directors. Unless there is some specific function in our movie application that needs to treat theatrical releases differently, there is no need to represent this distinction in our model.

How do I know that we mean the same thing by "film"? I will know through your action. If you suggest that all films can be nominated for an Academy Award, and *Sarah T.* is in my list of films, I will know that we don't mean the same thing by "film."

An ontology provides a precise vocabulary with which knowledge can be represented. This vocabulary allows us to specify which entities will be represented, how they can be grouped, and what relationships connect them together.

The vocabulary can be seen as a social contract between a data provider and a data consumer. The more precise the ontology, the greater the potential understanding of how the data can be used. But if the ontology is overly complex, including categories and relationships that a data consumer doesn't require, then it can become confusing, complicated, and difficult to use, maintain, and extend.

So how do we know if our ontology is the right one? We know it's right if the people and systems using it are able to interoperate successfully. This is the same standard that we apply to computer software. Does the program run? Do we get the right answer? If so, we are successful.

Models Are Data

Given that RDF is a meta-model, it is probably no surprise that an ontology can be expressed as RDF triples and stored in a graph store alongside the data it describes. It is important to recognize that these semantic models are simply another form of data that has no intrinsic behavior itself. It is only when software that understands semantic models operates on this data that interesting things happen.

In Chapter 3 we defined inference as a process that derives new information from existing information. We can think of the new information being generated as the conclusions we draw from the existing information. Up to this point, the rules for our inference processes have been built into our code. But ideally, we would like to expose the rules that drive our inference process so that others can write software that draws the same conclusions when working with our data.

Ontologies allow us to express the formal rules for inference. When software reads our ontology, it should have all the information necessary to draw the same conclusions from our data that we did. When the actions of software are consistent with the rules expressed in the ontology, we say that the software has made an *ontological commitment*.

In object-oriented (OO) programming, it is easy to forget how many rules the system is handling for us. For example, a subclass in an OO system contains its parent's members (methods and variables) because, somewhere in the system, software asserted that having access to parent members is consistent with what it means to be an OO subclass.

By design, semantic systems leave it up to you to build or invoke the necessary software to make inferences based on the model. Producing consistent actions based on data is quite literally what we mean by "semantic programming." As we will see, there are many semantic platforms and tools that can aid development by providing basic inferencing capabilities for common ontologies.

An Introduction to Data Modeling

We've already done plenty of data modeling in this book—you've seen examples of how to represent relationships between movies and actors; between restaurants, cuisine, and location; and many others. In each case, we've talked about these different classes of things and what properties these things have—for instance, movies have names, release dates, and actors, whereas restaurants have cuisines and locations. Classes are used to define the characteristics of a group of things, such as movies or actors, and to specify their relationships to other classes. In this section we'll show you how to build a type system that formally defines which properties are associated with which classes, and allows you to more easily say what something is.

Classes and Properties

Like object-oriented models, semantic models use the term *class* to describe groups of entities. In most OO systems, we say that an object is an instance of a class because it was constructed from the class definition—that is, the class served as a prototype or blueprint for the instance, and once the instance is created, it contains all the members (methods and variables) of the class.

Semantic data, as we have seen, is focused on the relationships between entities. So not surprisingly, semantic models are property-oriented rather than entity- (or object-) oriented. Thus, semantic entities aren't "born" into a class; rather, they are understood to be members of a class because of their properties. This may seem like a subtle distinction, but as we will see, this focus on relationships gives semantic models a great deal of flexibility.

In object-oriented systems, properties are defined as a part of a class. When you know which class an object is derived from, you know which properties it will contain. In semantic systems, properties are defined independently. A property definition may optionally indicate which types of resources have the property. The collection of types that use the property is called the *domain* of the property. And likewise, a property definition may also indicate which types of values this property can take on, representing the *range* of the property. See Figure 6-1.

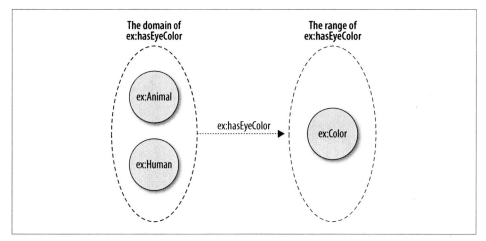

Figure 6-1. The domain and range of a property expressing eye color

When a property does not indicate its domain, we cannot infer anything about the resources it is describing, as it may be used by any type of resource. On the other hand, if the property does define a type as its domain, then we can infer that anything described by that property is of the domain type. Moreover, if the property defines several types as its domain, we can infer that the resource described by that property is *all* of the domain types.

Similarly, if a property does not specify a range, then we can't infer anything about the value of the property. But if the property specifies one or more types as its range, then a value of the property can be inferred to be all of the types specified by the property range.

Let's work through a simple model to highlight the differences between semantic and OO approaches to class and property definition. Consider a trivial model of a person that provides information about eye color. In an object-oriented system, we could define an `Animal` class with a `Human` subclass, which has a variable representing eye color. When we construct an instance of the `Human` class to represent Jamie, the instance would contain a variable to hold his eye color, which we could set to "blue." We would know that Jamie is both an animal and a human because the object representing him was constructed from the `Human` class.

In a fictional semantic model (using the `example.org` domain, which we will give the prefix ex), we could declare that the property `hasEyeColor` has a domain of `Animal` and `Human` with a range of `Color`:

```
ex:hasEyeColor rdf:type    rdf:Property .
ex:hasEyeColor rdfs:domain ex:Animal .
ex:hasEyeColor rdfs:domain ex:Human .
ex:hasEyeColor rdfs:range  ex:Color .
```

and then say that Jamie has blue eyes with the statement:

```
<http://semprog.com/people/jamie> ex:hasEyeColor
    <http://rdf.freebase.com/ns/en.blue> .
```

Based on this statement and the definition of `hasEyeColor`, it could be inferred that Jamie is an animal and a human and that blue is a color, even if no further assertions were made about either Jamie or blue.

Because semantic classes are defined on properties, it is also possible to define classes in terms of the value of a property. For instance, we could define a class that consists of people who have blue eyes. We need not assert that Jamie is a member of the class, since his membership could be inferred based on the assertion that he has blue eyes.

Of course, we can always explicitly assert that a resource is a member of a class. In fact, as we saw in Chapter 4 while examining RDF serializations, this happens so frequently that most RDF serializations have developed a shorthand for expressing it. But while we frequently assert class membership, it is important to understand that we can infer class membership through several mechanisms within semantic models.

Semantic Pseudocode and the example.org Domain

In RFC 2606, the Internet Engineering Task Force (IETF) set aside the domains `example.com`, `example.org`, and `example.net` for use as placeholders in documentation and examples where a domain name is required.

You will frequently see semantic "pseudocode" using resources with example-domain URIrefs. As with the previous examples, these resources don't actually exist, they simply serve as placeholders to show how something could be stated.

Modeling Films

Let's now build a data model for our movie dataset in Chapter 2. In that dataset, we have three different kinds of things that are related—movies, actors, and directors—and here we'll create a class for each one. Actors have film performances in films, and directors direct films, and everything has a name. Figure 6-2 displays a graph for the film *Blade Runner* showing that Ridley Scott is the director and Harrison Ford is an actor. For IDs and types we'll use the sp namespace, which corresponds to http://www.semprog.com/film#. Notice that we're using the rdf:type property to indicate what type of class something is. We're going to quickly introduce a lot of new vocabularies that use RDF, RDFS, and OWL. Don't be intimidated—we'll go into the specifics later in this chapter.

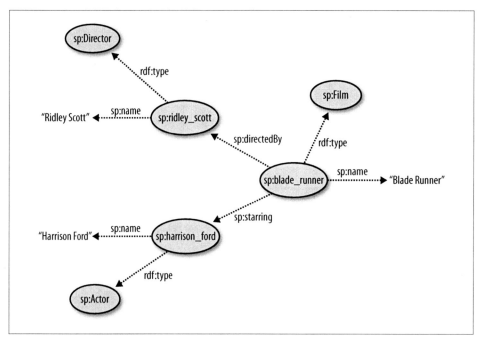

Figure 6-2. A graph describing the film Blade Runner

As you can see in Figure 6-2, the class sp:Film uses the properties sp:starring and sp:directedBy, and all three classes use the property sp:name. Now we're going to create the schema that describes the classes and properties. We'll start by defining a simple class hierarchy. In order to reuse the name property across multiple classes, we'll attach it to a class called sp:Object and then have other classes be subclasses of object. We'll also add in an sp:Person class that is a superclass of sp:actor and sp:director, so that we can later add properties common to people, such as birth dates and family relationships. We're using the rdfs:subclassOf property to connect subclasses to superclasses. See Figure 6-3.

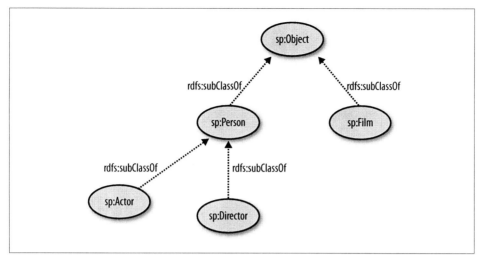

Figure 6-3. The class hierarchy for our film ontology

In addition to the class hierarchy, we need to declare the range and domain of the properties that we're using. As mentioned earlier, the range is the source type for a property, and the domain is the destination type for a property. The graph in Figure 6-4 shows the triples for defining the properties and types for our film data. You can see that we declare that `sp:Object`, `sp:Film`, and `sp:Director` are of type `owl:Class`; that `sp:starring`, and `sp:directedBy` are of type `owl:ObjectProperty`; and that `sp:name` is of type `owl:DatatypeProperty`. This is because `sp:name` points to a literal value, and the other properties point to resources. Also, the domains and ranges for the properties are defined. The property `sp:starring` has domain `sp:Film` and range `sp:Actor`, and `sp:directedBy` has domain `sp:Film` and range `sp:Director`.

The property `sp:name` is defined in a different way because name points to a string value. The range of name is `xsd:string`, which is an XML literal datatype. All XML literal datatypes such as floats, integers, strings, and dates can be used in OWL as long as the property that uses them is of type `owl:DatatypeProperty`.

Property and Class Naming in Ontologies

When you have a property or class whose name is composed of multiple words, the overwhelming consensus is to use camel-casing. This means that instead of using `directed_by` or `directedby`, you should used `directedBy` to name the "directed by" property. The first letter of class names is usually capitalized, and the first letter of property names is lowercase.

There isn't a consensus for other instance resources, as these identifiers are often system-dependent. However, short and readable identifiers are always better than purely machine-generated codes.

As the graph in Figure 6-4 suggests, defining classes and properties can become very complex very quickly. This is a place where tools and APIs can be a huge help.

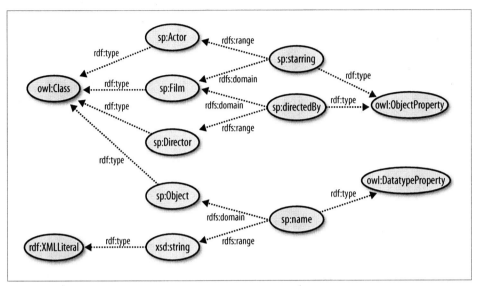

Figure 6-4. The property definitions for our film ontology

Reifying Relationships

A common problem in data modeling occurs when it is necessary to make statements about relationships. An example of this in our film dataset would be adding the role that an actor played in a movie. So, instead of saying that Harrison Ford was in the movie *Blade Runner*, we want to say that Harrison Ford played Rick Deckard in the movie *Blade Runner*. We can't simply add a role property because the role isn't a property of an actor or a film—it's a property of Harrison Ford's performance in *Blade Runner*.

The process of making a subject-predicate-object statement into a subject is called *reification* in RDF. The RDF standard has a vocabulary for reifying triples, but it's overly verbose and complicated and doesn't interoperate well with existing RDF APIs or with OWL or RDF Schema (RDFS). Instead, we'll take the more straightforward route of having the `sp:starring` property point to a class called `sp:Performance` that has an `sp:actor` property and an `sp:role` property. With this simple change, we can now make more statements about an actor's performance. Figure 6-5 shows our new schema.

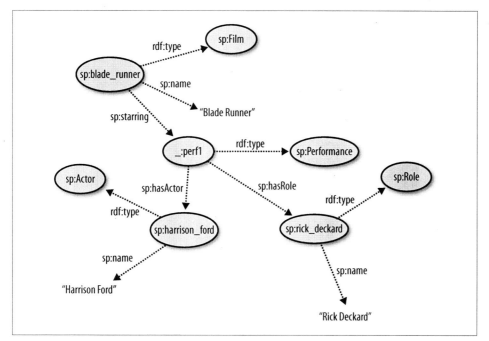

Figure 6-5. The performance class in the film ontology

The `sp:Performance` class is different from the `sp:Actor`, `sp:Film`, and `sp:Director` classes because, instead of describing a primary concept or thing in our data, it describes a relationship between things. Without an `sp:Actor`, `sp:Film`, or `sp:Role` connected to it, a specific `sp:Performance` is meaningless—it's dependent on things it is connected to in order to express meaning. In RDF, it is common to use blank nodes for reified relationships, as they may not have an intrinsic ID that comes from the data being modeled.

The pattern of reifying relationships is common in semantic data modeling. It adds expressiveness at the cost of greater complexity, so you probably don't want to reify all relationships, but don't be afraid to follow the reification pattern when your data demands it.

Just Enough OWL

The Web Ontology Language (OWL) is an RDF language developed by the W3C for defining classes and properties, and also for enabling more powerful reasoning and inference over relationships. OWL was built as an extension to RDFS, an earlier, simpler schema vocabulary, and is based on a lot of previous work developing ontology languages. OWL is the current W3C standard for defining semantic web schemas, and tools and API support for OWL are rapidly expanding.

OWL is a very large language with a lot of complicated parts. OWL itself is broken into three sub-languages of increasing complexity and expressiveness called OWL-Lite (the simplest), OWL DL, and OWL Full. We're going to cover just enough of OWL-Lite to be able to define types and properties, as these are the most useful applications of OWL.

The full vocabulary of OWL uses URIs in the RDF, RDFS, and OWL namespaces, and it also makes use of the XML Schema literal definitions. Following are the most important classes for creating an ontology:

owl:Thing
> The class of all things in OWL. All classes implicitly subclass this class, and all individual OWL instances are implicitly instances of owl:Thing. The use of this class is similar to the superclass Object in some object-oriented programming languages. Properties with a range or domain of rdfs:Resource can be used with any instance.

owl:Class
> The class of RDF resources that are classes. All classes are instances of rdfs:Class.

owl:DatatypeProperty
> The class of all properties that have ranges that are literals—and instances of rdfs:Datatype.

owl:ObjectProperty
> The class of all properties that have ranges that are instances of owl:Class.

rdf:XMLLiteral
> The class of all XML literal values, defined in the XML Schema spec. Generally, most ontologies use the XML Schema literal values exclusively, as they are well defined and well understood. This is a subclass of rdfs:Literal and an instance of rdfs:Datatype.

Along with those classes, the following are the most useful properties for creating an ontology:

rdf:type
> The type of a resource. This property specifies that a resource is an instance of a specific RDFS class.

rdfs:subClassOf
> Specifies that a class is a subclass of another class, and therefore that all instances of the subclass are also instances of the superclass.

rdfs:domain
> Specifies that a property has a domain of a specified class. In practice, this means that when the specified property is used in a triple, the subject of the triple will always be an instance of the class specified by the rdfs:domain property in the ontology.

`rdfs:range`

> Specifies that a property has a range of a specified class. In practice, this means that when the specified property is used in a triple, the object of the triple will always be an instance of the class specified by the `rdfs:range` property in the ontology. This can be used to specify literal range properties as well.

Here's the code for defining our earlier film schema in OWL using RDFLib:

```
from rdflib import ConjunctiveGraph, Namespace, Literal, BNode

# define the OWL, RDF, RDFS, and XSD URIs:
owlNS = Namespace("http://www.w3.org/2002/07/owl#")
owlClass = owlNS["Class"]
owlObjectProperty = owlNS["ObjectProperty"]
owlDatatypeProperty = owlNS["DatatypeProperty"]
rdfNS = Namespace("http://www.w3.org/1999/02/22-rdf-syntax-ns#")Ob
rdfProperty = rdfNS["Property"]
rdfType = rdfNS["type"]
rdfsNS = Namespace("http://www.w3.org/2000/01/rdf-schema#")
rdfsSubClassOf = rdfsNS["subClassOf"]
rdfsDomain = rdfsNS["domain"]
rdfsRange = rdfsNS["range"]
xsdNS = Namespace("http://www.w3.org/2001/XMLSchema#")
xsdString = xsdNS["string"]

# define the namespace and classes:
filmNS = Namespace("http://www.semprog.com/film#")
objectClass = filmNS['Object']
personClass = filmNS['Person']
filmClass = filmNS['Film']
performanceClass = filmNS['Performance']
actorClass = filmNS['Actor']
roleClass = filmNS['Role']
directorClass = filmNS['Director']
# define the properties:
name = filmNS['name']
starring = filmNS['starring']
hasActor = filmNS['has_actor']
hasRole = filmNS['has_role']
directedBy = filmNS['directed_by']

# define all of our triples:
schemaTriples = [from rdflib import ConjunctiveGraph, Namespace, Literal

# define the OWL, RDF, RDFS, and XSD URIs:
owlNS = Namespace("http://www.w3.org/2002/07/owl#")
owlClass = owlNS["Class"]
owlObjectProperty = owlNS["ObjectProperty"]
owlDatatypeProperty = owlNS["DatatypeProperty"]
rdfNS = Namespace("http://www.w3.org/1999/02/22-rdf-syntax-ns#")
rdfProperty = rdfNS["Property"]
rdfType = rdfNS["type"]
rdfsNS = Namespace("http://www.w3.org/2000/01/rdf-schema#")
rdfsSubClassOf = rdfsNS["subClassOf"]
rdfsDomain = rdfsNS["domain"]
```

```
rdfsRange = rdfsNS["range"]
xsdNS = Namespace("http://www.w3.org/2001/XMLSchema#")
xsdString = xsdNS["string"]

# define our namespace:
filmNS = Namespace("http://www.semprog.com/film#")

# define all of our triples:
schemaTriples = [
    # class declarations:
    (filmNS['Object'], rdfType, owlClass),
    (filmNS['Person'], rdfType, owlClass),
    (filmNS['Film'], rdfType, owlClass),
    (filmNS['Actor'], rdfType, owlClass),
    (filmNS['Role'], rdfType, owlClass),
    (filmNS['Director'], rdfType, owlClass),
    (filmNS['Performance'], rdfType, owlClass),
    # class heirarchy:
    (filmNS['Film'], rdfsSubClassOf, filmNS['Object']),
    (filmNS['Person'], rdfsSubClassOf, filmNS['Object']),
    (filmNS['Actor'], rdfsSubClassOf, filmNS['Person']),
    (filmNS['Role'], rdfsSubClassOf, filmNS['Object']),
    (filmNS['Director'], rdfsSubClassOf, filmNS['Person']),
    # name property:
    (filmNS['name'], rdfType, owlDatatypeProperty),
    (filmNS['name'], rdfsDomain, filmNS['Object']),
    (filmNS['name'], rdfsRange, xsdString),
    # starring property:
    (filmNS['starring'], rdfType, owlObjectProperty),
    (filmNS['starring'], rdfsDomain, filmNS['Film']),
    (filmNS['starring'], rdfsRange, filmNS['Performance']),
    # hasActor property:
    (filmNS['hasActor'], rdfType, owlObjectProperty),
    (filmNS['hasActor'], rdfsDomain, filmNS['Performance']),
    (filmNS['hasActor'], rdfsRange, filmNS['Actor']),
    # hasRole property:
    (filmNS['hasRole'], rdfType, owlObjectProperty),
    (filmNS['hasRole'], rdfsDomain, filmNS['Performance']),
    (filmNS['hasRole'], rdfsRange, filmNS['Role']),
    # directedBy property:
    (filmNS['directedBy'], rdfType, owlObjectProperty),
    (filmNS['directedBy'], rdfsDomain, filmNS['Film']),
    (filmNS['directedBy'], rdfsRange, filmNS['Director'])
]

# create a graph and add the triples:
graph = ConjunctiveGraph()
for triple in schemaTriples: graph.add(triple)
```

Because we're storing the schema in a graph, we can query it and get a list of classes and properties:

```
>>> list(graph.subjects(rdfType, owlClass))
[rdflib.URIRef('http://www.semprog.com/film#Role'),
    rdflib.URIRef('http://www.semprog.com/film#Film'),
    rdflib.URIRef('http://www.semprog.com/film#Actor'),
```

```
    rdflib.URIRef('http://www.semprog.com/film#Person'),
    rdflib.URIRef('http://www.semprog.com/film#Director'),
    rdflib.URIRef('http://www.semprog.com/film#Object')]
>>> list(graph.subjects(rdfType, owlObjectProperty))
[rdflib.URIRef('http://www.semprog.com/film#directedBy'),
    rdflib.URIRef('http://www.semprog.com/film#hasRole'),
    rdflib.URIRef('http://www.semprog.com/film#hasActor'),
    rdflib.URIRef('http://www.semprog.com/film#starring')]
>>> list(graph.subjects(rdfType, owlDatatypeProperty))
[rdflib.URIRef('http://www.semprog.com/film#name')]
```

A common operation when working with classes is to check whether one class is equal to or a subclass of another class. We do this by doing a depth-first search up the class hierarchy:

```
def isSubClassOf(subClass, superClass, graph):
    if subClass == superClass: return True
    for parentClass in graph.objects(subClass, rdfsSubClassOf):
        if isSubClassOf(parentClass, superClass, graph): return True
        else: return False
```

Then we can check which classes are people:

```
>>> isSubClassOf(filmNS['Actor'], filmNS['Person'], graph)
True
>>> isSubClassOf(filmNS['Film'], filmNS['Person'], graph)
False
```

Next, we'll add our film data about *Blade Runner*. Notice that we're defining resources for the film, actor, director, role, and performance:

```
# define a blank node for the performance:
performance = BNode('_:perf1')

# define the triples:
filmTriples = [
    # movie:
    (filmNS['blade_runner'], rdfType, filmNS['Film']),
    (filmNS['blade_runner'], filmNS['name'],
        Literal("Blade Runner", datatype=xsdString)),
    (filmNS['blade_runner'], filmNS['starring'], performance),
    (filmNS['blade_runner'], filmNS['directedBy'], filmNS['ridley_scott']),
    # performance:
    (performance, rdfType, filmNS['Performance']),
    (performance, filmNS['hasActor'], filmNS['harrison_ford']),
    (performance, filmNS['hasRole'], filmNS['rick_deckard']),
    # actor:
    (filmNS['harrison_ford'], rdfType, filmNS['Actor']),
    (filmNS['harrison_ford'], filmNS['name'],
        Literal("Harrison Ford", datatype=xsdString)),
    # role:
    (filmNS['rick_deckard'], rdfType, filmNS['Role']),
    (filmNS['rick_deckard'], filmNS['name'],
        Literal("Rick Deckard", datatype=xsdString)),
```

```
    # director:
    (filmNS['ridley_scott'], rdfType, filmNS['Director']),
    (filmNS['ridley_scott'], filmNS['name'],
        Literal("Ridley Scott", datatype=xsdString))
]

for triple in filmTriples: graph.add(triple)
```

RDF schemas are usually stored in the same graph as the data that the schema describes. This makes it easy to query the data and schema together. Here's how to find every resource in the graph that is an instance of a type by walking the class hierarchy:

```
def findInstances(queryClass, graph, instances=None):
    if instances is None: instances = set()
    for instance in graph.subjects(rdfType, queryClass): instances.add(instance)
    for subClass in graph.subjects(rdfsSubClassOf, queryClass):
        findInstances(subClass, graph, instances)
    return instances
```

Now we can ask for all instances of person:

```
>>> findInstances(personClass, graph)
set([rdflib.URIRef('http://www.semprog.com/film/ridley_scott'),
    rdflib.URIRef('http://www.semprog.com/film/harrison_ford')])
```

Using Protégé

Developing OWL ontologies using just triples is too difficult to do for large ontologies. Luckily, there are a number of tools for building and maintaining large OWL ontologies. We'll cover Protégé, which is one of the most heavily developed and feature-rich tools available for OWL. Protégé is an open source Java GUI developed at Stanford University; releases are available at *http://protege.stanford.edu/*. We'll be using Protégé 4.0 in these examples, so download and install the correct version for your computer in order to follow along. If you're using an earlier version, such as Protégé 3.4, this walk-through may not make sense.

Creating a New Ontology

When you run Protégé, you'll be greeted with a menu asking whether you want to create a new ontology or edit an existing one. Select "Create new OWL ontology," and Protégé will ask you for your ontology base URI, to which all RDF URIs will be relative. For our film ontology, we're using the base URI http://www.semprog.com/film#. For your own ontologies, you should use a URI that corresponds to the project and the organization that the project is for. See Figure 6-6.

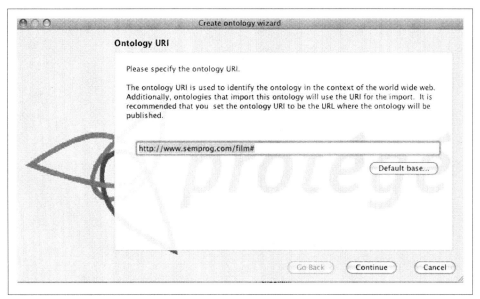

Figure 6-6. Creating an ontology in Protégé

Next, Protégé will prompt you for a physical location for your ontology. Choose where you want the project files to be saved, and then select Finish to go to the ontology editor.

Editing an Ontology

The first thing we need to do is build the class hierarchy. Select the Classes tab, and then select the class Thing. This is the root class in OWL—all classes subclass it. Press the "Add subclass" button with the "+" on it to add new subclasses. Figure 6-7 shows the hierarchy for our film ontology; build it out by adding the individual subclasses.

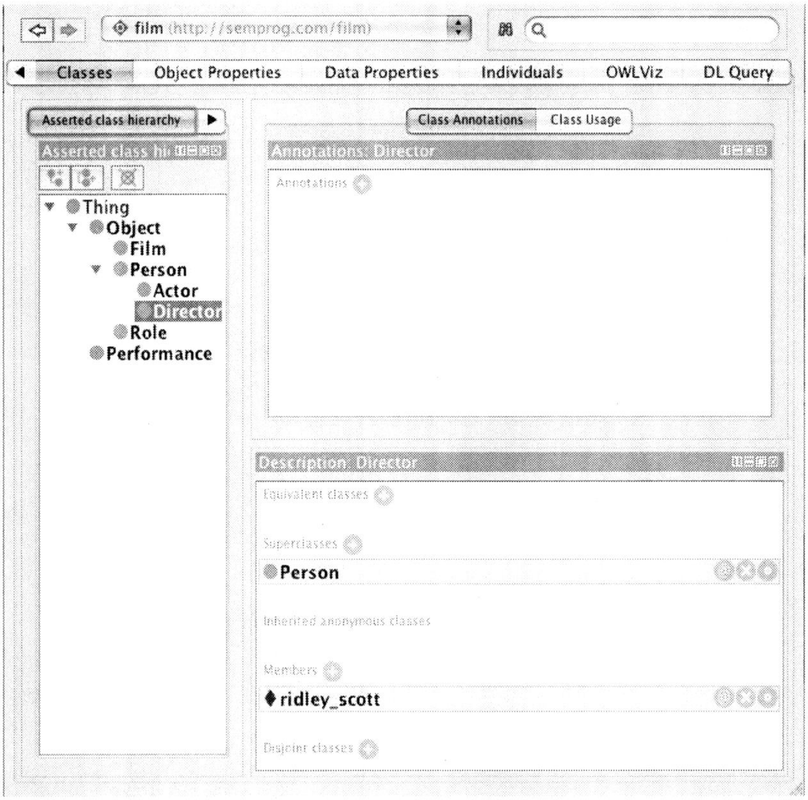

Figure 6-7. Creating the class hierarchy in Protégé

Next, we need to add the properties. OWL distinguishes between object properties that point to objects and datatype properties that point to literal values. Select the "Object Properties" tab, and then press the "Add property" button with the "+" on it to add a new property. After you add a property, you can set the domain (the source class of the property) and the range (the destination class of the property) by pressing the "+" button next to "Domains" and "Ranges." In the domain and range dialogs, you can type in the class name or choose the class from the hierarchy.

Figure 6-8 shows the filled-out properties. We've entered the domains and ranges for each property as well.

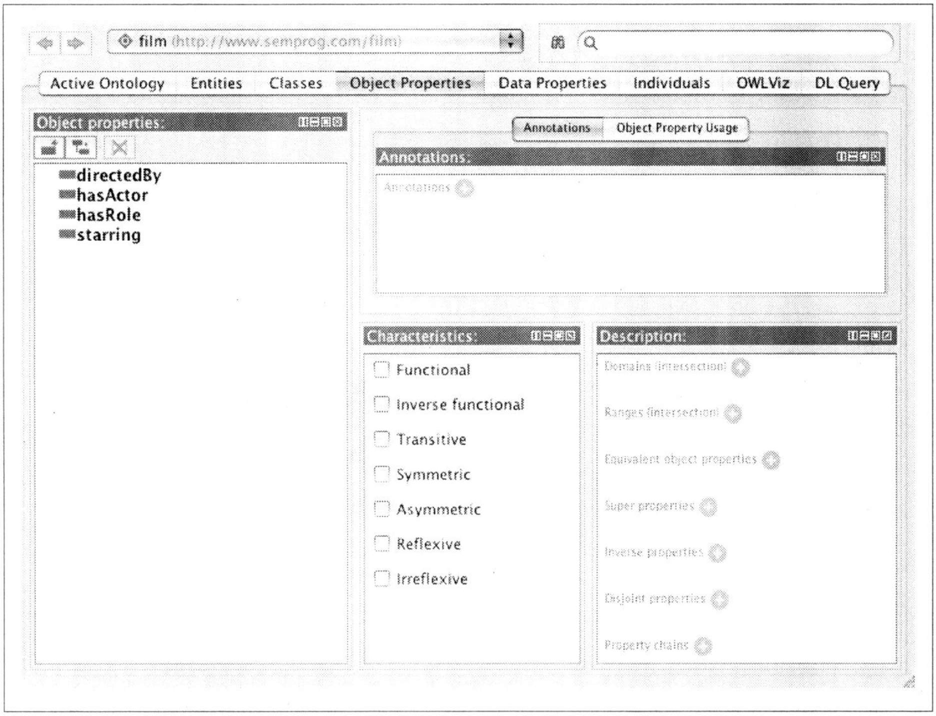

Figure 6-8. Creating object properties in Protégé

Finally, we need to describe the name property. It's a datatype property, so go to the "Data Properties" tab, and then press the "Add property" button with the "+" in order to add it. Enter the domain by pressing the "+" button next to "Domains" and selecting a class. Select the range by pressing the "+" next to "Ranges," and select the "string" literal type from the list of types that Protégé knows about. Figure 6-9 shows how the tab looks.

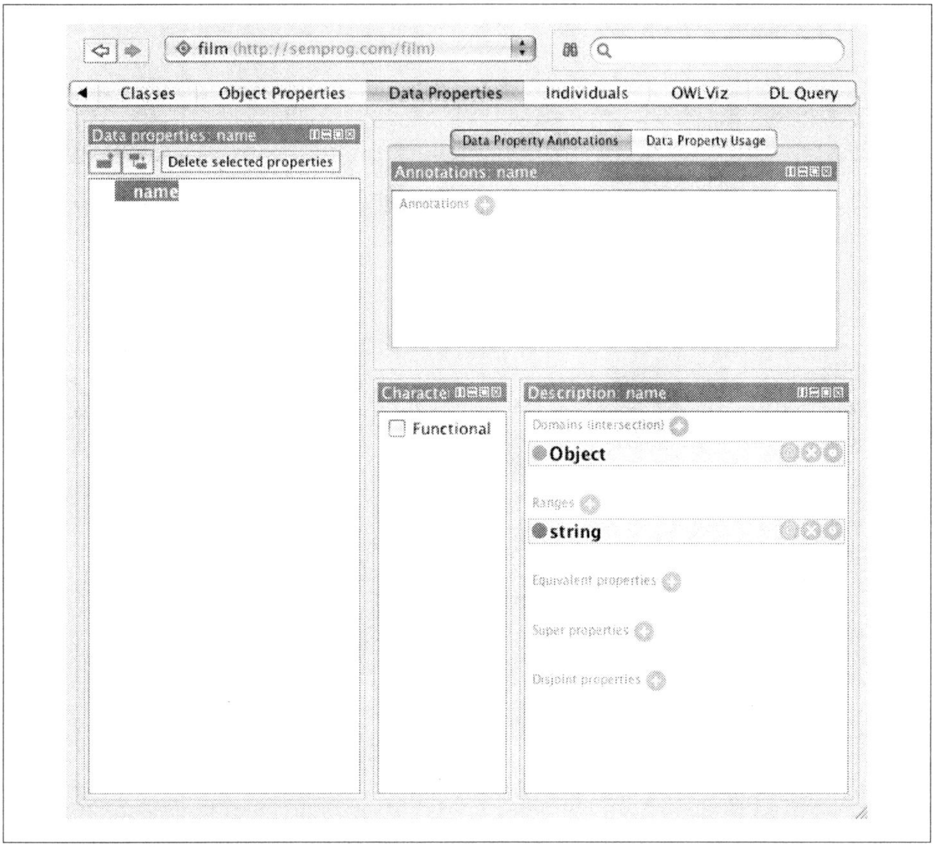

Figure 6-9. Creating data properties in Protégé

Protégé has support for creating instances of your classes once the ontology is defined. Generally, you'll want to load most of your data through scripts and programs instead of by hand, but the Individuals tab allows you to test out your ontology by filling in records. You'll need to add the type and property values for each instance. Figure 6-10 shows our *Blade Runner* data from earlier.

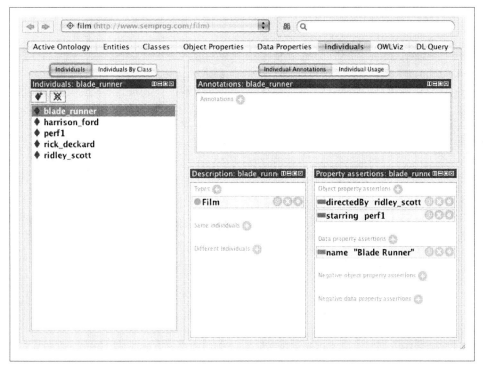

Figure 6-10. Creating individual instances in Protégé

You've just completed your first ontology! The OWL ontology is saved by Protégé as an RDF/XML file, and can be loaded into RDFLib easily:

```
from rdflib import ConjunctiveGraph
graph = ConjunctiveGraph()
graph.load("film-ontology.owl")
```

Just a Bit More OWL

While working with Protégé you may have noticed some optional information you can provide on classes and properties. These modifiers allow you to remove ambiguities about how things relate to one another.

Functional and Inverse Functional Properties

Marking a property as functional means that there is only one object for a given subject with the property. For instance, in our film ontology, `has_actor` and `has_role` would be functional properties, because for each individual performance, there can only be a single actor and a single role. Marking a property as inverse functional has the inverse meaning: there is only one subject for a given predicate and object. We could mark `starring` as an inverse functional property, because each performance can only be in one movie.

The classes `owl:FunctionalProperty` and `owl:InverseFunctionalProperty` are subclasses of `owl:ObjectProperty`, so you declare a property to be functional or inverse functional by specifying the type of the property using `rdf:type`.

Inverse Properties

RDF statements imply a direction on the relationship, going from the domain of the subject to the range of the object. Thus we say that films are directed by directors, but the converse—directors direct films—is not specifically implied by our model. Rather than modeling an independent second set of assertions that reciprocate the film-director relationship, OWL allows us to state:

```
<http://www.semprog.com/film#directed> owl:inverseOf
    <http://www.semprog.com/film#directedBy> .
```

Disjoint Classes

Members of a class that is disjoint from a second class are guaranteed not to be members of the second class. For instance, if we wanted to differentiate made-for-TV movies from theatrical movies, we could define a `TVMovie` class as disjoint from our `Film` class. But doing so would also preclude our `TVMovie` class from being a subclass of our `Film` class (since instances of a subclass are by definition instances of the parent class). This may not be a good choice, since we would then need to define properties like director and performance separately for the `TVMovie` class. See Figure 6-11.

More usefully, we could define `TVMovie` as a subclass of `Film` and create another class called `TheatricalMovie` to represent movies appearing in theaters. This would also subclass our `Film` class and specify that the `TheatricalMovie` class is disjoint from the `TVMovie` class. See Figure 6-12.

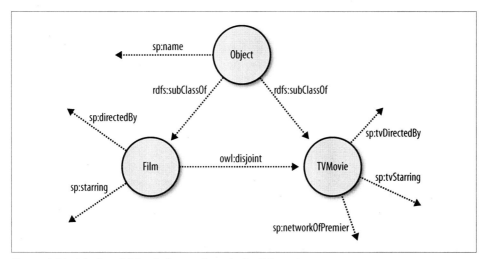

Figure 6-11. A bad model for dealing with made-for-TV movies

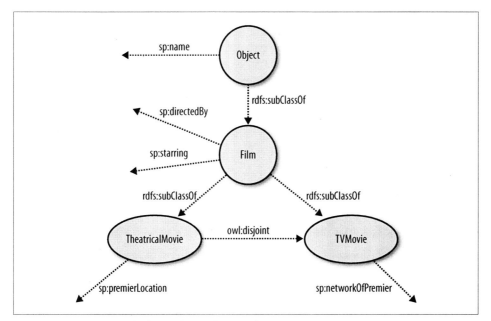

Figure 6-12. A better model for dealing with made-for-TV movies

Keepin' It Real

When we develop an ontology (or a model), we are explicitly selecting to describe the aspects of the world that are important to our application. Decisions about what to exclude from our ontology are as important as decisions about what to include. A good ontology is like a good map, providing enough information to keep you from getting lost, but never more than you need to find your way.

It is important to remain pragmatic when developing ontologies—they are not ends in themselves, but are simply frameworks for expression. If no one encodes knowledge using a particular ontology, or if no application is written that does something useful with the knowledge, then the value of the ontology is zero.

You may recall that in Chapter 4 we were able to query our movie data without being aware of any formal ontology. In fact, for the sake of simplicity, our movie data did not make any `rdf:type` assertions, and yet our SPARQL queries worked perfectly well. While this is not necessarily good practice, the point is that for many purposes, a simple understanding of the data, without significant regard for the ontological model, can be "good enough."

It is also interesting to point out that the instance data and definitions in semantic systems are represented in the same way and frequently exist together in the same repository. Metadata is data. This means you can inspect metadata in the same way you inspect instance data, and you can query both simultaneously. Rather than having a data dictionary or DDL that is built, maintained, and used separately from the data itself, the data description is a part of the data itself. As we have discussed previously, this is a critical part of what makes semantic data portable and reusable.

Some Other Ontologies

Learning to build effective data models is a skill that comes from experience. Examining other data models and playing with instance data using the model is a great way to learn new data modeling patterns and vicariously absorb the wisdom of other modelers.

A good place to find example models is the Stanford Protégé group's wiki of ontologies (*http://protegewiki.stanford.edu/index.php/Protege_Ontology_Library#OWL_on tologies*). Many of the models assume a deep understanding of the subject matter domain, but others, such as the Space Shuttle Crew ontology, demonstrate basic concepts that generalize to any employment/project domain.

Describing FOAF

We have used and discussed FOAF informally, but there is a formal OWL definition of the vocabulary. The latest version of the FOAF vocabulary is described (in human-readable form) at *http://xmlns.com/foaf/spec/#term_homepage*. And you can find the latest RDF (OWL) description of the ontology at *http://xmlns.com/foaf/spec/index.rdf*.

It's worth taking the time to look through the ontology, as you will discover that FOAF is being used for much more than social networks. Because it provides predicates for describing web pages, photographs, organizations, computer programs, and more, FOAF has become one of the most used (and incorporated) vocabularies in the world.

A Beer Ontology

Reading through ontologies isn't much fun, and although Protégé can facilitate your understanding of an ontology, having a picture makes it even easier to grasp an ontology's overall design. Happily, if you have the Graphviz program installed, Protégé can produce pictures of ontologies much like the ones we have shown in this chapter.

Graphviz is a graph visualization package from AT&T, distributed under a public license, that produces images of graphs expressed in the DOT file format. You may recall from Chapter 3 that we used the pydot package to produce DOT visualizations of celebrity relationships. Graphviz can be downloaded from *http://www.graphviz.org* and is also available through most package managers.

Install the version of Graphviz appropriate to your platform (Protégé on the Mac can use the MacPorts distribution or the source build, but it has problems with the *Graphviz.app* package) and start Protégé. Under the Options or Preferences menu item (depending on your platform), enter the path to the *dot* application in the panel labeled OWLViz. See Figure 6-13.

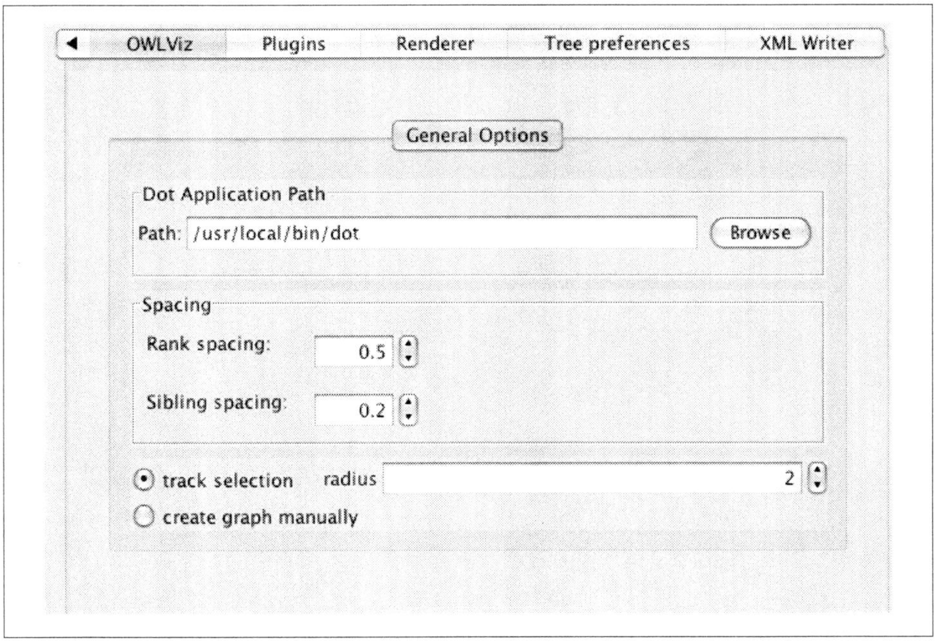

Figure 6-13. Setting the location of Graphviz in Protégé

Protégé can load OWL files directly from a URI, so under the File menu select "Open from URI..." and grab a copy of the beer ontology from *http://www.dayf.de/2004/owl/ beer.owl*. You can now explore the ontology using the tabs we used while creating the movie ontology. For instance, clicking on the Classes tab and expanding the Thing class will display the top-level classes in the ontology. See Figure 6-14.

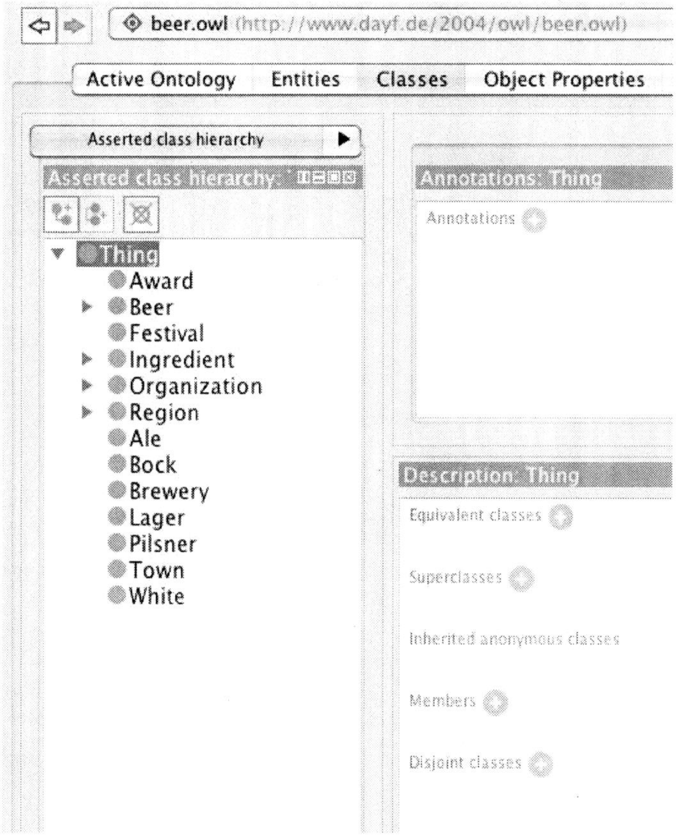

Figure 6-14. Beer ontology top-level classes in Protégé

To view the ontology graphically, click on the OWLViz tab and select the Thing object from the asserted class hierarchy tab on the left side. You should now see the first two levels of the ontology's class hierarchy. See Figure 6-15.

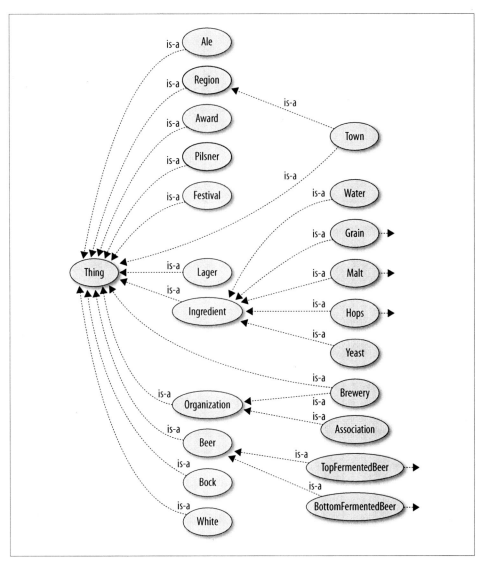

Figure 6-15. Protégé's rendering of the beer ontology

Protégé allows you a good deal of control over what is rendered graphically and how it is displayed. Using the toolbar above the graph, you can select different rendering options and expand different components of the ontology. To focus in on an area, use the class hierarchy on the left side to select the classes of interest.

Protégé is a great way to learn, view, and modify ontologies. As you explore ontologies in Protégé, try populating them with some instance data and saving the graph to a local file. By loading the graph into your RDF library, you can run queries against the data to get a better feel for how ontologies affect query patterns.

This Is Not My Beautiful Relational Schema!

When building a data model, remember that no one gets it right the first time. Be happy if it seems like you are getting closer the third time around. Data modeling is not a singular activity, it is iterative: build a model, populate some instances, run some queries, repeat.

The good news is that semantic data modeling is very different from relational data modeling. Assertions between entities can be made at any time. If tomorrow you discover a new type of relationship between entities that increases the utility of your data, you can simply go ahead and assert the relationship and update the ontology to reflect the change. There is no need to bring the data store down, and no schema reorganization needs to take place. When you add the assertions, they will be available for use on the next query. Old queries will continue to work, and new queries that make use of the assertion will be enhanced.

It's also important to remember that unlike a relational data model, a semantic data model is not a monolithic thing. Good semantic data models borrow vocabularies whenever possible from other ontologies. A semantic data model is a collection of ontologies (or vocabularies) that work together to represent the slice of the world you are focusing on. Start with a broad model that frames the problem space and provides basic organization to your entities. Then, look around for other ontologies that help to describe the entities and represent the more complex relationships they have with one another. The more you borrow, the more portable and extensible your data becomes. If you find yourself working on an ontology that seems to be "all encompassing," stop and think about how you can break it down into modules and where you might be able to incorporate other vocabularies into those modules.

Also, because semantic models define properties in terms of the things they connect, semantic models are extremely open-ended. Anyone at any time can define a new property that connects entities. A property definition may come from a source that had no involvement in the creation or publication of the class definitions. In this way semantic models are also highly collaborative, allowing communities to borrow and extend models with little oversight or coordination.

Some of this flexibility is hard to believe if you are steeped in the world of relational databases—especially in a high-availability production environment. One way to start experiencing the power of semantic data stores is to write a small application that lets you track some aspect of your life: your workout schedule, data about a hobby or sport, or some aspect of family life. As you start to use the simple app, you will think of other, related things you could track. Try expanding the data model to incorporate those relationships and slowly extend the application to make use of the new relationships.

In Chapters 9 and 10 we will build larger, more complex applications, but you should now have enough capabilities in your mental toolbox to build a few small things yourself.

Not the Last Word on Ontology

We have only begun to scratch the surface of semantic modeling in this chapter. Although it provides a useful starting point, modeling is best learned by examining a large collection of well-built models.

For the next step in your journey through semantic modeling, we recommend the book *Semantic Web for the Working Ontologist* by Dean Allemang and James Hendler (Morgan Kaufmann). These authors take a more complete look at RDFS and inferencing, and they explore the most useful pieces of OWL. Their combinations-and-patterns approach to teaching semantic modeling provides you with a strong foundation for taking your modeling capabilities to the next level.

## CHAPTER 7

Publishing Semantic Data

Publishing semantic data often results in a "network effect" because of the connections that occur between datasets. For example, publishing restaurant data can suddenly open up a new application for a connected geographic dataset. Campaign finance data is made more accessible by its connections to politicians and their voting records. As more semantic data is published, existing semantic data becomes more useful, which in turn increases the scope and usefulness of the applications that can be built.

The main barrier to publishing semantic data has been that many of the standards are complicated and confusing, and there has been a lack of good tools. Fortunately, there have been several efforts to create simpler semantic web standards that make it easier for both designers and developers to publish their data. Another promising trend is that thousands of web applications now have open APIs and are publishing machine-readable data with implicit semantics that can easily be translated to the explicit semantics that we've been covering. All of this has led to early efforts by some very large web properties to consume semantic data and use it to enhance their applications.

In this chapter we'll look at two emerging standards, Microformats and RDFa, that aim to make publishing semantic data far easier for web designers by simply extending the already familiar HTML syntax. We'll also look at ways to take data that you might already have access to—either through existing online APIs or in spreadsheets and relational databases—put that data into an RDF store, and publish it using the same Linked Open Data techniques you saw in Chapter 5. After reading this chapter, we hope that you'll be both prepared and inspired to publish explicit semantics more often in your future work, making the sharing and remixing of data easier for everyone.

Embedding Semantics

One of the major criticisms of semantic web formats like RDF/XML is that they are too complicated and too much of a hassle for a designer or webmaster to bother implementing. Furthermore, a huge amount of information is already available on HTML pages, and duplicating it in a different format is both a significant upfront investment and an ongoing maintenance hassle.

Microformats and RDFa work to address these issues by allowing users to embed semantic tags in existing web pages. By making a few small changes to a web page, the semantics of the hyperlinks and information on the page become explicit, and web crawlers can more accurately extract meaning from the page. We'll look at both of these standards here and consider their strengths and drawbacks.

Microformats

Microformats are intended to be a very simple way for web developers to add semantic data to their pages by using the familiar HTML `class` attribute. The website *http://microformats.org* defines a variety of microformats, both stable and under development. To understand how they work, take a look at this simple example of an hCard, a business-card-like microformat for giving information about a person or organization that can be embedded in HTML:

```
<div class="vcard">
  <div class="fn">Toby Segaran</div>
  <div class="org">The Semantic Programmers</div>
  <div class="tel">919-555-1234</div>
  <a class="url" href="http://kiwitobes.com/">http://kiwitobes.com/</a>
</div>
```

---

hCard or vCard?

The vCard data interchange format is defined in RFC 2426 and is used by many address book applications (including the Address Book application that ships with Mac OS X). The hCard format provides a simple mapping of vCard elements to HTML elements. Confusingly, the outer element of an hCard representation uses vcard as its class name.

The first `div` uses the `class` attribute to tell us that it contains a `vcard`. The hCard specification (which you can find at *http://microformats.org*) defines several fields, including:

fn
> Formatted name

org
> Organization

tel
> Telephone number

url
> Person's web page

The HTML elements within the `vcard` div have classes matching these properties. The values for the properties are the text inside the tags, so the hCard shown earlier is describing a person whose full name is "Toby Segaran" and who works for "The Semantic Programmers". By taking an existing HTML web page and adding microformat annotations, you make the page more interpretable by a machine. For instance, a web crawler could use microformats found on a website to build a database of people and their phone numbers, employers, and web pages.

The earlier hCard example is just a list of fields that would appear on a web page as a small table or box, but most of the interesting information on web pages is contained within natural language text. In many cases a fragment of text makes reference to a single thing such as a place, a person, or an event, and microformats can be used to add semantics to these references. Here is an example of using the "hCalendar event" microformat to add semantics to a sentence:

```
<p class="vevent">
    The <span class="summary">English Wikipedia was launched</span>
    on 15 January 2001 with a party from
    <abbr class="dtstart" title="2001-01-15T14:00:00+06:00">2</abbr>-
    <abbr class="dtend" title="2001-01-15T16:00:00+06:00">4</abbr>pm at
    <span class="location">Jimmy Wales' house</span>
    (<a class="url" href=
        "http://en.wikipedia.org/wiki/History_of_Wikipedia">more information</a>)
</p>
```

hCalendar or Event?

As hCard is to vCard, hCalendar is to iCalendar (the data interchange format used by Apple's iCal and defined in RFC 2445). Events are actually subproperties of the hCalendar specification, but common practice allows you to omit the wrapper class (vcalendar) if you are simply identifying events.

This example shows the use of the summary, dtstart, dtend, location, and url properties from the vEvent microformat, which are added to an existing sentence to capture some of the semantics of what the sentence says. It also illustrates the case where the information we want to show differs from the information we want to capture. Notice how the displayed start time within the dtstart is simply 2. This is sufficient for display but not sufficient for a parser, so it is overridden by the title attribute, which gives the complete time in ISO 8601 date format.

Again, by using this microformat you can add semantic data to a page and make it possible for crawlers to create more interesting applications. In this case, historical data about past events could be republished as a timeline or made searchable by the time-frame of the events described.

Here's one more example of a microformat that you should probably be using on your own website! It's the hResume microformat, which allows you to describe your work and educational experience. It is particularly interesting because it nests the hCalendar and hCard microformats inside it, showing how microformats can build on each other:

```
<div class="hresume">
...
<li class="experience vevent vcard">
```

```
<object  data="#name" class="include"></object>

<h4 class="org summary">
  <a href="http://www.metaweb.com" >
    Metaweb Technologies
  </a>
</h4>
<p class="organization-details">1000-5000 employees</p>
<p class="period">
    <abbr class="dtstart" title="2003-06-01">January 2008</abbr>
&mdash;<abbr class="dtend" title="2005-08-01">December 2009</abbr>
    <abbr class="duration" title="P2Y3M">(1 year 11 months)</abbr>
</p>
<p class="description">
   Designed and implemented large-scale data-reconciliation techniques.
</p>
</li>
...
</div>
```

This example shows one experience line item extracted from a larger hResume div.
Notice how the li tag has a class attribute with experience (from hResume), vEvent
(from the hCalendar microformat), and vCard (from the hCard microformat). Besides
the fact that there are fewer class names to remember, mixing formats like this means
that a crawler that isn't aware of the hResume format could still determine that there
was an organization called "Metaweb" and know its URL if it understood the hCard
microformat.

LinkedIn, the largest professionally focused social network, embeds the hResume
microformat in its public profile pages.

RDFa

We described RDFa in Chapter 4 when we covered RDF serializations, but RDFa is
really a mechanism for publishing semantic data within standard web pages. Like
microformats, it works by adding attributes to tags that define fields. However, it also
allows anyone to define a namespace in the same way that RDF/XML does. Because of
this, publishers aren't restricted to officially sanctioned vocabularies, and they can
define their own if nothing appropriate already exists.

The attributes used for RDFa are different from those used for microformats, but they
serve similar functions. Here (again) are some of the more important attributes defined
by RDFa:

about
    A URI (or safe CURIE) used as a subject in an RDF triple. By default, the base URI
    for the page is the root URI for all statements. Using an about attribute allows
    statements to be made where the base URI isn't the subject.

rel
> CURIEs expressing relationships between two resources.

rev
> CURIEs expressing reverse relationships between two resources.

property
> CURIEs expressing relationships between a resource and a literal.

src
> A URI resource expressing an RDF object (as an inline embedded item).

content
> A string, representing a literal RDF object.

href
> A URI resource expressing an RDF object (as inline clickable).

resource
> A URI (or safe CURIE) expressing an RDF object when the object isn't visible on the page.

datatype
> The datatype of a literal.

typeof
> The type of a subject.

The namespace being used is specified in the same way as in RDF/XML, XHTML, or any XML-based format, by using xmlns attributes. CURIEs are a superset of XML QNames, so in the examples that follow we will simply use QName constructions where CURIEs are required.

Here's a simple example of using the familiar FOAF namespace to embed some semantic information into HTML:

```
<body xmlns:dc="http://purl.org/dc/elements/1.1/"
      xmlns:foaf="http://xmlns.com/foaf/0.1/">

  <h1>Toby's Home Page</h1>
  <p>My name is
    <span property="foaf:firstname">Toby</span> and my
    <span rel="foaf:interest" resource="urn:ISBN:0752820907">favorite
    book</span> is the inspiring <span about="urn:ISBN:0752820907"><cite
    property="dc:title">Weaving the Web</cite> by
    <span property="dc:creator">Tim Berners-Lee</span></span>
  </p>
</body>
```

A sentence about Toby has been marked up with semantics specifying that he's a person with the first name "Toby" and that he likes a book called "Weaving the Web", which has the creator "Tim Berners-Lee".

The body tag contains two XML namespaces—FOAF, specified by the xmlns:foaf attribute, and the Dublin Core namespace, specified by xmlns:dc, the same way these were declared in Chapter 5. The property attribute uses the defined namespace to indicate what the contents of a tag mean, and the resource attribute is used to specify a unique identifier for the tag. Thus, we know exactly which "Toby" it is and which book we're referring to.

Here is a simpler example that uses the beer namespace, which you can find at *http://www.purl.org/net/ontology/beer*:

```
<div xmlns:beer="http://www.purl.org/net/ontology/beer#">
  <div about="#Guiness" typeof="beer:Stout">
    <span property="beer:hasAlcoholicContent">7.5</span>% Alcohol
  </div>
</div>
```

RDFa is relatively easy to add to existing web pages, whether they're dynamic or static. At the time of this writing, there are more large services publishing microformats than RDFa, which is a newer standard. However, services that consume embedded semantics (which we'll get to in a moment) are striving for compatibility with both. If you want to use semantics that aren't supported by an existing microformat, you'll have to use RDFa. Since it's best if everyone is using the same schemas, check out *http://www.schemaweb.info/* to see if someone has already created a schema for your application.

Yahoo! SearchMonkey

Services that consume embedded semantic data are already starting to appear. One of the earliest examples is SearchMonkey, an effort by Yahoo! to integrate structured and semantic data into Yahoo!'s search results. At the time of this writing, it's still very much under development, but we believe that it gives a first hint of what's possible when publishers provide even a small amount of structured data along with their human-readable pages.

There are two parts to SearchMonkey. As the Yahoo! web crawler indexes websites, SearchMonkey extracts structured data from the pages. Then, the SearchMonkey APIs allow developers to create SearchMonkey applications that format structured data into richer, more useful search results. A few of these applications are already part of Yahoo!'s main search results—Figure 7-1 shows an example of a result I get when I search for "dosa mission san francisco" in Yahoo! search.

Figure 7-1. The Citysearch application for SearchMonkey

For structured data, SearchMonkey currently supports several standards and approaches, including RDFa, microformats, an Atom-based update feed, and the OpenSearch API. All of these methods deliver structured metadata about web pages on your website when it is being indexed by Yahoo!. Without this structured data, Yahoo!'s crawling and indexing process has to analyze the text and links on your web pages to try to discern what your website is about and whether it is a useful result to someone's search query. By using a data schema and adding structure to the pages on your site, you're making it easier for Yahoo! to understand your content, and Yahoo! can then do a better job of displaying your website as a search result.

Creating a Yahoo! SearchMonkey application generally involves a small amount of PHP, which is hosted by Yahoo!. We won't go into the details here, but you can learn more at *http://developer.yahoo.com/searchmonkey/*. Figure 7-2 shows a few examples of applications that developers have already created and placed in the Yahoo! Search Gallery, which users can add to enrich their search results.

Figure 7-2. Examples of SearchMonkey applications in the Yahoo! Search Gallery

Google's Rich Snippets

Google has also started indexing semantic data expressed as RDFa. Like SearchMonkey, Google's initial use of RDFa metadata is to locate information on the page that can be used to enrich search results. To guide content developers, Google has published a small vocabulary covering frequently searched subjects such as people, organizations, products, and reviews.

When a content creator embeds RDFa that uses the *http://rdf.data-vocabulary.org/rdf .xml* vocabulary on a page, Google's search system will use the RDFa to locate specific pieces of content to display in query results. For instance, by identifying where on the page the product review rating is located, along with markup indicating the price and the person writing the review, CNET's product reviews deliver summary information to users much faster than similar, nonsemantically enabled results returned by the same query. See Figure 7-3.

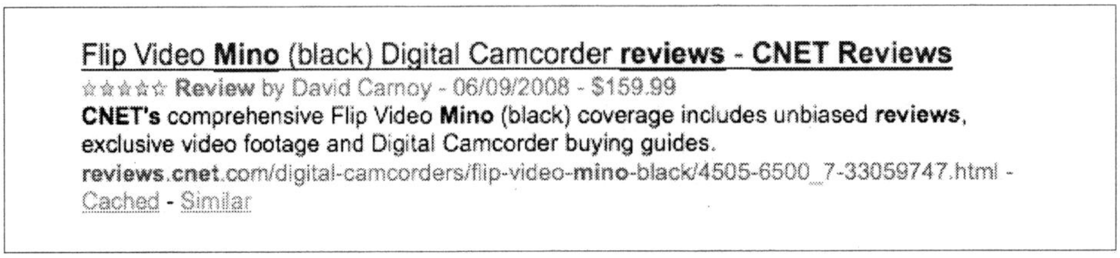

Flip Video Mino (black) Digital Camcorder **reviews - CNET Reviews**
☆☆☆☆☆ **Review** by David Carnoy - 06/09/2008 - $159.99
CNET's comprehensive Flip Video **Mino** (black) coverage includes unbiased **reviews**, exclusive video footage and Digital Camcorder buying guides.
reviews.**cnet**.com/digital-camcorders/flip-video-**mino**-black/4505-6500_7-33059747.html - Cached - Similar

Figure 7-3. Examples of SearchMonkey applications in the Yahoo! Search Gallery

Dealing with Legacy Data

While it's great that data is increasingly available on the Web in standard formats, it's also true that the vast majority of data that most people have access to—both on the Web and in their private data store—is not. In this section we'll look at a few examples covering open APIs, web pages, and relational databases, to help you get some idea of how you might incorporate different kinds of data into your RDF triplestore and be able to publish it in one of the semantic web formats we've discussed.

Although we'll take you through code for getting data from a few different places, the purpose of this section is really to show you a basic pattern for taking data from anywhere and converting it to RDF, so you can both store it and republish it for consumption by other applications. This generally breaks down into three steps:

1. Identify and parse a source of data.
2. Find or create a schema that matches your data.
3. Map the data to that namespace and make some RDF.

Simple enough, right? Now let's try it out on some real datasets.

Internet Video Archive

The Internet Video Archive (IVA) is an aggregator of video content, such as movie trailers and video game previews, that you can find at *http://internetvideoarchive.com*. It provides an API that allows developers to include video clips on their sites and that also gives lots of information about movies, music, and TV shows. To get started, you'll need to get an API key by going to *http://api.internetvideoarchive.com/*. This should take only a few seconds, but if you'd rather not bother getting an API key, we've provided static versions of the files on *http://semprog.com*.

The Movies API exposes several methods, which are accessible through simple REST requests. You can search for movies by name, look for what's new on DVD, or, as we'll do in this case, look at data about movies currently in theaters. To see how this API call works, point your web browser to *http://www.videodetective.com/api/intheaters .aspx?DeveloperId={YOUR KEY}* (or *http://semprog.com/data/intheatres.xml*) and you should see something like this:

```
<items Page="1" PageSize="100" PageCount="1" TotalRecordCount="69">
  <item>
    <Description>QUANTUM OF SOLACE Video</Description>
    <Title>QUANTUM OF SOLACE</Title>
    <Studio>Columbia Pictures</Studio>
    <StudioID>255</StudioID>
    <Rating>PG-13</Rating>

    <Genre>Action-Adventure</Genre>
    <GenreID>1</GenreID>
    <Warning>Sex, violence</Warning>
    <ReleaseDate>Sat, 01 Nov 2008 00:00:00 GMT</ReleaseDate>
    <Director>Marc Forster</Director>
    <DirectorID>18952</DirectorID>

    <Actor1>Giancarlo Giannini</Actor1>
    <ActorId1>2929</ActorId1>
    <Actor2>Judi Dench</Actor2>
    <ActorId2>4446</ActorId2>
    <Actor3>Daniel Craig</Actor3>
    <ActorId3>14053</ActorId3>
    ...
    <PublishedId>629936</PublishedId>
    <Link>http://www.videodetective.com/titledetails.aspx?publishedid=808130</Link>
    <Duration>143</Duration>
    <DateCreated>Fri, 11 Apr 2008 13:31:00 GMT</DateCreated>
    <DateModified>Wed, 29 Oct 2008 11:04:00 GMT</DateModified>
    <Image>http://www.videodetective.com/photos/1263/05308032_.jpg</Image>
    <EmbedCode>
        <![CDATA[<embed src="http://www.videodetective.net/flash/apimovieplayer.swf"
        ...
    </EmbedCode>
    ...
  </item>
</items>
```

This is a pretty standard XML response to a REST query. It contains a list of items, each of which has a set of fields describing each item. In this case, the fields are information about a movie, such as the title, director, actors, and genre. It also contains IDs for the movie itself and for all the people involved.

Now that we've identified the source, we need to choose or create a namespace for the fields that we want to map the data into. Ideally, we would choose a standard namespace that a lot of other people are using so that we could merge our datasets with theirs. While there's no official standard for what to call the different aspects of a movie, the Freebase movie schema (*http://www.freebase.com/type/schema/film/film*) is used in many contexts both within and outside Freebase. See Figure 7-4.

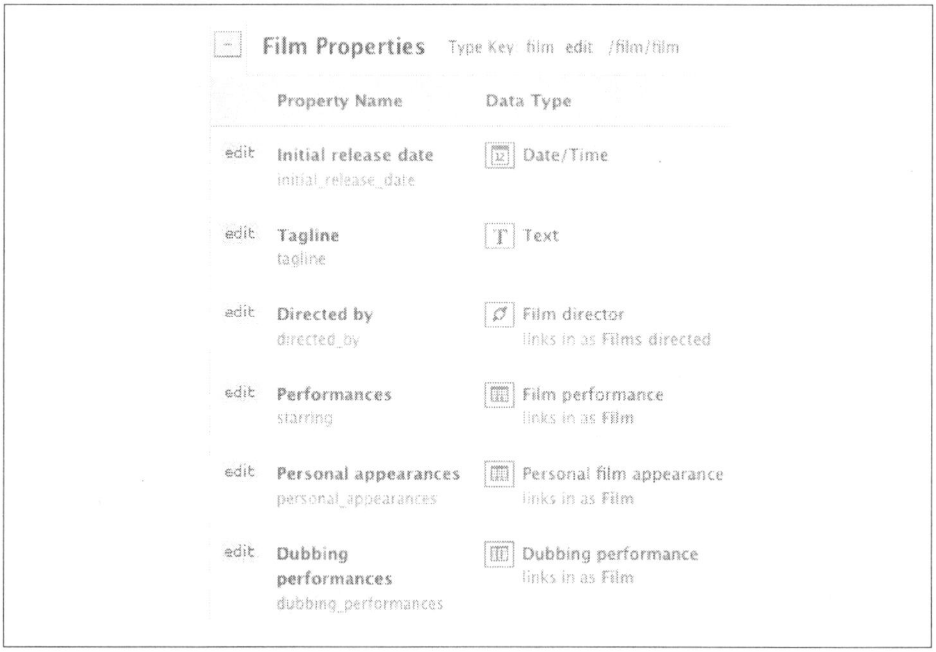

	Film Properties	Type Key film edit /film/film
	Property Name	**Data Type**
edit	**Initial release date** initial_release_date	🗓 Date/Time
edit	**Tagline** tagline	T Text
edit	**Directed by** directed_by	⌀ Film director links in as Films directed
edit	**Performances** starring	▦ Film performance links in as Film
edit	**Personal appearances** personal_appearances	▦ Personal film appearance links in as Film
edit	**Dubbing performances** dubbing_performances	▦ Dubbing performance links in as Film

Figure 7-4. Partial Freebase film schema

Generally, the fields in the film schema used by Freebase, which you can see in Figure 7-4, approximately match the fields in the XML file that we're getting from the IVA. There is one notable exception: the Freebase schema has `Performances` instead of `Actors`. If you drill down into the definition, you'll see that a performance links to both an `Actor` and a `Role`, which describes the character that the actor plays in the film. In the case of IVA, the characters aren't given, so we'll leave `Role` empty for our example. Our goal is to translate the XML records into a graph that looks like what you see in Figure 7-5.

Now that we have the source and the schema, we can write the code to download and parse the XML, and then put it in an RDF store using our namespaces. Take a look at *IVAtoRDF.py*, which you can download from *http://semprog.com/psw/chapter7/IVAtoRDF.py*. The first section is just a few inserts and namespace definitions, a pattern you'll use again and again as you express new datasets as RDF:

```
from rdflib.Graph import ConjunctiveGraph
from rdflib import Namespace, BNode, Literal, RDF, URIRef
from urllib import urlopen
from xml.dom.minidom import parse

FB = Namespace("http://rdf.freebase.com/ns/")
IVA_MOVIE= Namespace("http://www.videodetective.com/titledetails.aspx?publishedid=")
IVA_PERSON= Namespace("http://www.videodetective.com/actordetails.aspx?performerid=")
RDFS = Namespace("http://www.w3.org/2000/01/rdf-schema#")
```

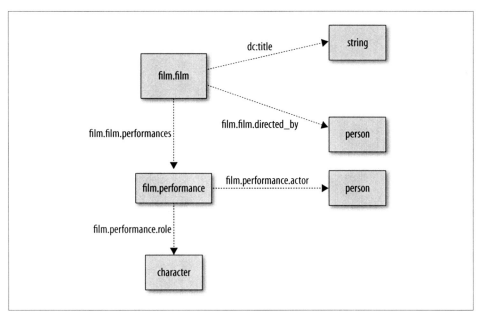

Figure 7-5. Target graph structure for film data (Freebase and Dublin Core)

In this code, we're using the Freebase and RDF namespaces. Since the IVA doesn't have an official schema for expressing their data in RDF, we've also invented some new namespaces to represent items coming from the IVA. We've imported parse from the minidom API that comes with Python to parse the XML.

The next piece of code defines a method that parses the XML from IVA into a list of dictionaries. Each dictionary will contain the movie ID, title, director, and actors. The director and actors are also dictionaries, each containing an ID and a name. You could extract even more information about the movie from the XML, but we'll keep it simple for this example. The main method here is get_in_theaters, which opens and parses the XML file and then uses standard DOM operations to get the relevant data:

```
# Returns the text inside the first element with this tag
def getdata(node,tag):
    datanode = node.getElementsByTagName(tag)[0]
    if not datanode.hasChildNodes(): return None
    return datanode.firstChild.data

# Creates a list of movies in theaters right now
def get_in_theaters():
    # use this if you have a key:
    #stream =
        urlopen('http://www.videodetective.com/api/intheaters.aspx?DeveloperId={KEY}')
    # otherwise use our copy of the data:
    stream = urlopen('http://semprog.com/data/intheatres.xml')
    root=parse(stream)
    stream.close()
```

```
movies=[]
for item in root.getElementsByTagName('item'):
    movie={}
    # Get the ID, title, and director
    movie['id'] = getdata(item,'PublishedId')
    movie['title'] = getdata(item,'Title')
    movie['director'] = {'id':getdata(item,'DirectorID'),
        'name':getdata(item,'Director')}
    # Actor tags are numbered: Actor1, Actor2, etc.
    movie['actors'] = []
    for i in range(1,6):
        actor = getdata(item,'Actor%d' % i)
        actorid = getdata(item,'ActorId%d' % i)
        if actor != None and actorid != None:
            movie['actors'].append({'name':actor, 'id':actorid})
    movies.append(movie)
return movies
```

Now here's the interesting part: we want to take the dictionary and express it as a graph like the one shown in Figure 7-5. To do this, we're going to loop over all the movies, construct a node representing each one, and add the literal properties, the ID and the title, to that node. Next we'll create a director node and link it to the movie. Finally (and this is the slightly tricky part), we'll loop over each actor and create an anonymous **performance** node and an **actor** node. The actor gets linked to the performance, and the performance gets linked to the movie.

If this seems overly complicated, remember that the **performance** node is there so that if we ever decide to add the name of the character to the actor, we'll have a concept of "performance"—that is, a node to which the role can be connected:

```
# Generate an RDF Graph from the Movie Data
def make_rdf_graph(movies):
    mg = ConjunctiveGraph()
    for movie in movies:
        # Make a movie node
        movie_node = IVA_MOVIE[movie['id']]
        mg.add((movie_node,DC['title'], Literal(movie['title'])))
        # Make the director node, give it a name and link it to the movie
        dir_node = IVA_PERSON[movie['director']['id']]
        mg.add((movie_node,FB['film.film.directed_by'], dir_node))
        mg.add((dir_node, DC['title'], Literal(movie['director']['name'])))
        for actor in movie['actors']:
            # The performance node is a blank node -- it has no URI
            performance = BNode()
            # The performance is connected to the actor and the movie
            actor_node = IVA_PERSON[actor['id']]
            mg.add((actor_node,DC['title'], Literal(actor['name'])))
            mg.add((performance, FB['film.performance.actor'], actor_node))
            # If you had the name of the role, you could also add it to the
            # performance node, e.g.
            # mg.add((performance, FB['film.performance.role'],
            #     Literal('Carrie Bradshaw')))
```

```
            mg.add((movie_node, FB['film.film.performances'], performance))
        return mg
```

Finally, we just need a main method to tie it all together. Since the IVA API has a lot of different options, we could replace get_in_theaters with another function that returns a dictionary of movies, and convert that to RDF instead:

```
if __name__=='__main__':
    movies = get_in_theaters()
    movie_graph = make_rdf_graph(movies)
    print movie_graph.serialize(format='xml')
```

What we've done here, and what you'll be seeing for the remainder of this section, is extracted the data from its source and recreated it with explicit semantics. Additionally, where possible, we've pointed to a published schema so that others who are using that schema can easily consume the data, and those who aren't can at least read what all the fields mean. As an exercise, see if you can find another published film schema. Then, figure out how many movies are available in RDF using that schema, and alter the code so it creates an RDF file using that namespace.

Tables and Spreadsheets

A huge amount of data, both on the Web and saved on people's computers, is in tabular formats—many people believe that there's more business data in Excel spreadsheets than in any other format. As we learned in Chapter 1, and as we've been seeing throughout this book, data in tables is easy to read but difficult to extend, and more importantly, it can't easily be combined with data from other sources.

To extend the previous example, consider the following CSV file, which was exported from Excel and can be downloaded from *http://semprog.com/psw/chapter7/MovieRe views.csv*. It's a pretty simple file, with just three fields: a movie name, a rating, and a short review. It matches the movies provided in the sample file in the previous section:

```
QUANTUM OF SOLACE,3,"This film will please action fans, but the franchise has all...
FILTH AND WISDOM,1,Madonna's directorial debut is unconvincing and incoherent
ROLE MODELS,4,"Juvenile, ridiculous and predictable, yet it still manages to be...
FEARS OF THE DARK,5,"This French animated horror portmanteau is monochrome and...
CHANGELING,2,What could have been a dramatic and tumultuous film ends up boring...
```

What we're going to do here demonstrates how semantic data can be combined from multiple sources using different schemas. In the previous section we used the Freebase movie schema, but Freebase contains mostly facts and doesn't have any fields for opinions or reviews. However, there's a standard RDF review vocabulary available at *http://www.purl.org/stuff/rev#* that we can use to attach the reviews in the file to the movies already in the graph. The review vocabulary is not specific to movies, and can be used for describing reviews of anything. Even though the Freebase schema that we used doesn't support movie reviews, we can add them simply by including another namespace. Even better, the namespace is a de facto standard and is used in hReview microformats, so there will certainly be other services able to read it.

The following code, which you can download from *http://semprog.com/psw/chapter7/ MergeTabReviews.py*, shows how to add the reviews to the existing graph. It first creates the movie graph using the functions defined earlier, then loops over the CSV file and searches for movies by name to find the appropriate node. For each movie, it creates a BNode representing the review, and then adds the rating and review text to the review node. The final addition to the node is a reference to Toby, who wrote the review:

```
from rdflib.Graph import ConjunctiveGraph
from rdflib import Namespace, BNode, Literal, RDF, URIRef
from IVAtoRDF import FB,DC,get_in_theaters,make_rdf_graph
from csv import reader

# Reviews Namespace
REV=Namespace('http://www.purl.org/stuff/rev#')

if __name__=='__main__':

    # Create a graph of movies currently in theaters
    movies = get_in_theaters()
    movie_graph = make_rdf_graph(movies)

    # Loop over all reviews in the CSV file
    for title, rating, review in reader(open('MovieReviews.csv', 'U')):

        # Find a movie with this title
        match = movie_graph.query('SELECT ?movie WHERE {?movie dc:title "%s" .}' \
            % title, initNs={'dc':DC})

        for movie_node, in match:
            # Create a blank review node
            review_node = BNode()

            # Connect the review to the movie
            movie_graph.add((movie_node, REV['hasReview'], review_node))

            # Connect the details of the review to the review node
            movie_graph.add((review_node, REV['rating'], Literal(int(rating))))
            movie_graph.add((review_node, DC['description'], Literal(review)))
            movie_graph.add((review_node, REV['reviewer'], \
                URIRef('http://semprog.com/people/toby')))


    # Search for movies that have a rating of 4 or higher and the directors
    res=movie_graph.query("""SELECT ?title ?rating ?dirname
                            WHERE {?m rev:hasReview ?rev .
                                   ?m dc:title ?title .
                                   ?m fb:film.film.directed_by ?d .
                                   ?d dc:title ?dirname .
                                   ?rev rev:rating ?rating .
                                   FILTER (?rating >= 4)
                                   }""", initNs={'rev':REV, 'dc':DC, 'fb':FB})

    for title,rating,dirname in res:
        print '%s\t%s\t%s' % (title, dirname, rating)
```

At the end of the code is a SPARQL query to find the movie title, director, and rating for everything with a rating of 4 or higher. Although in this case you only have one review for each movie, it's possible to attach multiple review nodes to each one. That way, the query would search for everything that *anyone* rated 4 or higher. Running this program gives the output:

```
$ python MergeTabReviews.py
FEARS OF THE DARK   Marie Caillou   5
ROLE MODELS         David Wain      4
```

Notice that we've merged data from two different sources using two different schemas, and we can use both in a single query. This was possible because the movies had the same names in both datasets. If there were spelling mistakes or different punctuation, we would have had to clean them up in both sources to make the query possible.

Semantic MediaWiki

Wikis are an increasingly popular means of managing community-generated content. Like spreadsheets, wikis are intuitive to read and extremely easy to author, but paradoxically, as a wiki grows, its utility may remain constant due to the challenges of managing loosely structured content. Users will frequently repeat the same information on many pages and fail to update all instances of the information during updates, making the wiki inconsistent and stale.

The Semantic MediaWiki (SMW) is an extension to the MediaWiki platform, the wiki that powers Wikipedia. Unlike traditional wikis, the Semantic MediaWiki allows users to specify how a piece of content on one page is related to content at other locations across the wiki. Not only do these relations help keep the wiki consistent and organized, but they also allow queries to be written that take advantage of the structure. Lists and tables need not be hand-curated, as in traditional wikis, but rather can be the product of semantic queries across the wiki's content.

To make things even more interesting, most Semantic MediaWikis provide a resource with the path */wiki/Special:ExportRDF*. This will generate RDF/XML output of any page on the wiki, turning a simple MediaWiki into a true semantic publishing platform. For instance, you see RDF describing the SMW project itself at *http://semantic-media wiki.org/wiki/Special:ExportRDF/SMW_Project*.

Learn more about the Semantic MediaWiki at *http://semantic-mediawiki.org/wiki/Se mantic_MediaWiki*.

Legacy Relational Data

Almost all business and web applications are built on relational databases that are generally queryable with SQL. There are a number of reasons that we've explored so far for why one might choose to expose this relational data in a standard semantic format like RDF. In addition to the extensibility advantages of representing data semantically, it also makes it easy to share with others.

To demonstrate the conversion of typical relational data into RDF, we'll consider an incredibly simple schema for a message board, shown in Figure 7-6. The database consists of four tables that contain messages, users, and subjects. They are connected in the standard way, with foreign keys and an intermediate table for the many-many relationship between subjects and messages.

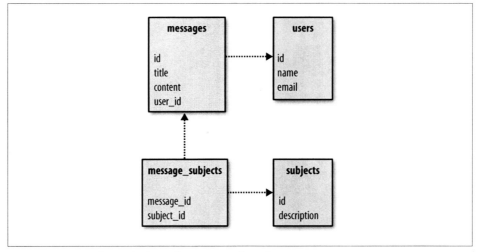

Figure 7-6. Basic relational schema for a message board

We've provided a SQL schema at *http://semprog.com/psw/chapter7/message_board .sql*. It creates the tables and inserts fake data into them. We had actually hoped to find a set of SQL data that you could just download, but there's very little out there. This perhaps demonstrates the point that SQL is a good tool for manipulating data but not as good for publishing and sharing data.

Our goal is to take this data and republish it as RDF using the Semantically-Interlinked Online Communities (SIOC, pronounced "shock") vocabulary, which is used to semantically describe conversations happening online. Messages published in SIOC can be more easily understood by aggregators, much like RSS feeds, but also have the ability to be connected across sites through the **has_reply** and **reply_of** properties. This means that it's possible to find the threads of conversations that are decentralized and published across the Web.

The code to do the conversion is available at *http://semprog.com/psw/chapter7/message _board_to_sioc.py*. Let's look through it so you can learn the basics of how this conversion works. The first part of the file defines all the namespaces we'll be using. SIOC and several others are used, including FOAF, since that's the standard way of expressing email addresses. We also throw in a namespace for this message board:

```
import sqlite3,os
from rdflib.Graph import ConjunctiveGraph
from rdflib import Namespace, BNode, Literal, RDF, URIRef
```

```
from urllib import urlopen

SIOC=Namespace('http://rdfs.org/sioc/ns#')
DC = Namespace("http://purl.org/dc/elements/1.1/")
DCTERMS = Namespace('http://purl.org/dc/terms/')
FOAF = Namespace("http://xmlns.com/foaf/0.1/")
RDFS = Namespace("http://www.w3.org/2000/01/rdf-schema#")

# Fake namespace for this message board
MB = Namespace('http://messageboard.com/')
```

The load_data function simply creates a sqlite database and runs all the SQL statements in the input file. This will get run by our main method as follows if it can't find a message_board database. Usually you would be starting with an already populated database, not a series of SQL statements, but this is the easiest way for us to distribute the data:

```
# load the SQL file into a database
def load_data(sqlfile, dbfile):
    conn = sqlite3.connect(dbfile)
    cur = conn.cursor()
    f = file(sqlfile)
    for line in f: cur.execute(line)
    f.close()
    conn.commit()
    conn.close()
```

The function that actually queries the database and creates the RDF graph is message_board_to_sioc. It's very simple—all it does is query the tables one at a time and create the relevant nodes. Except for in the message_subjects table, each row represents a node in a graph. Because we're using unique names in namespaces to refer to nodes, it's possible to refer to them before creating all their details, which means we don't have to do any joins! Just select from the table and create nodes in the graph:

```
# convert the message board SQL database to SIOC
def message_board_to_sioc(dbfile):
    sg = ConjunctiveGraph()
    sg.bind('foaf', FOAF)
    sg.bind('sioc', SIOC)
    sg.bind('dc', DC)

    conn = sqlite3.connect(dbfile)
    cur = conn.cursor()

    # Get all the messages and add them to the graph
    cur.execute('SELECT id, title, content, user FROM messages')

    for id, title, content, user in cur.fetchall():
        mnode = MB['messages/%d' % id]
        sg.add((mnode, RDF.type, SIOC['Post']))
        sg.add((mnode, DC['title'], Literal(title)))
        sg.add((mnode, SIOC['content'], Literal(content)))
        sg.add((mnode, SIOC['has_creator'], MB['users/%s' % user]))
```

```
# Get all the users and add them to the graph
cur.execute('SELECT id,name,email FROM users')
for id, name, email in cur.fetchall():
    sg.add((mnode, RDF.type,SIOC['User']))
    unode = MB['users/%d' % id]
    sg.add((unode, FOAF['name'], Literal(name)))
    sg.add((unode, FOAF['email'], Literal(email)))

# Get subjects
cur.execute('SELECT id,description FROM subjects')
for id, description in cur.fetchall():
    sg.add((mnode, RDF.type,DCTERMS['subject']))
    sg.add((MB['subjects/%d' % id], RDFS['label'], Literal(description)))

# Link subject to messages
cur.execute('SELECT message_id,subject_id FROM message_subjects')
for mid, sid in cur.fetchall():
    sg.add((MB['messages/%s' % mid],SIOC['topic'], MB['subjects/%s'] % sid))

conn.close()
return sg
```

Of course, this is a very simple database, and most relational databases will probably have many more tables. The important thing to understand is that this code takes a SQL database, which has implicit semantics, and converts it to an RDF graph, which has explicit semantics. It does this by taking the implied connections between tables (e.g., a user_id of 1 in a row in the messages table), and restating them in a standard vocabulary (e.g., triples like (message:10 has_creator user:1), where has_creator is defined by SIOC).

Finally, the main method ties it all together:

```
if __name__=="__main__":
    if not os.path.exists('message_board.db'):
        load_data('message_board.sql', 'message_board.db')
    sg = message_board_to_sioc('message_board.db')
    print sg.serialize(format='xml')
```

If you run message_board_to_sioc, you should see the RDF/XML version of the messages. Plug-ins are available for many of the popular blogging and message board platforms, which you can find at *http://sioc-project.org*.

RDFLib to Linked Data

In the previous section, you saw how to take datasets stored or published in other formats and load them into RDFLib, then generate all the XML for the graph you built. In practice, if you were consuming a very large graph with hundreds of thousands of nodes, you probably wouldn't want to publish the graph as a single XML file. In this section, we'll look at a more practical way to take a graph that you've built in RDFLib and turn it into a set of files that could be easily consumed by a crawler.

In Chapter 5 you learned about the Linking Open Data community, which has developed a set of conventions for building graph datasets that are distributed across the Internet. You can use the same ideas to republish a large graph as a series of files, which could then be served from one or more web servers.

The following is a simple example of how to do this with the data collected from the Internet Video Archive. You can download this example from *http://semprog.com/psw/chapter7/publishedLinkedMovies.py*:

```
from rdflib.Graph import ConjunctiveGraph
from rdflib import Namespace, BNode, Literal, RDF, URIRef
import IVAtoRDF

FB = Namespace("http://rdf.freebase.com/ns/")
DC = Namespace("http://purl.org/dc/elements/1.1/")
IVA_MOVIE= Namespace("http://www.videodetective.com/titledetails.aspx?publishedid=")
IVA_PERSON= Namespace("http://www.videodetective.com/actordetails.aspx?performerid=")
RDFS = Namespace("http://www.w3.org/2000/01/rdf-schema#")

movies = IVAtoRDF.get_in_theaters()
movie_graph = IVAtoRDF.make_rdf_graph(movies)

fq = movie_graph.query("""SELECT ?film ?act ?perf ?an ?fn WHERE
                        {?film fb:film.film.performances ?perf .
                         ?perf fb:film.performance.actor ?act .
                         ?act dc:title ?an.
                         ?film dc:title ?fn .
                         }""",
                    initNs={'fb':FB,'dc':DC})

graphs={}
for film, act, perf, an, fn in fq:
    filmid = fn.split(',')[0].replace(' ','_') + '_' + str(film).split('=')[1]
    actid = an.replace(' ','_') + '_' + str(act).split('=')[1]

    graphs.setdefault(filmid, ConjunctiveGraph())
    graphs.setdefault(actid, ConjunctiveGraph())

    graphs[filmid].add((film, FB['film.film.performance.actor'], act))
    graphs[filmid].add((act, OWL['sameAs'], actid + '.xml'))
    graphs[filmid].add((film, DC['title'], fn))

    graphs[actid].add((act, FB['film.actor.performance.film'], film))
    graphs[actid].add((film, OWL['sameAs'], filmid + '.xml'))
    graphs[actid].add((act, DC['title'], an))

for id, graph in graphs.items():
    graph.serialize('open_films/%s.xml' % id)
```

This code first calls a couple of methods from `IVAtoRDF` to generate the movie graph. It then generates a set of files for the movies and another set of files for the actors, in each case using a SPARQL query to pull out all the important data about the particular entity of interest. The data is amended with `owl:sameAs` links, which refer to the other files in the set, and each entity is saved in a small graph that is in turn saved to a small RDF/ XML file. The names of the XML files include the full names of the actors and movies to make things easier for search engines, and they also include the IDs to ensure uniqueness if two movies have the same name.

Executing this code will create a series of XML files like this:

```
Bill_Murray_761.xml
BODY_OF_LIES_39065.xml
BREAKFAST_WITH_SCOT_48168.xml
Brigette_Lin_38717.xml
```

For example, if you look at *BODY_OF_LIES_39065.xml*, you can see that the `owl:sameAs` links refer to other "information resources," so that a crawler knows where to go to find out more about each entity referred to in the file:

```
<?xml version="1.0" encoding="UTF-8"?>
<rdf:RDF
    xmlns:_3="http://rdf.freebase.com/ns/"
    xmlns:_4="http://purl.org/dc/elements/1.1/"
    xmlns:rdf="http://www.w3.org/1999/02/22-rdf-syntax-ns#"
    xmlns:rdfs="http://www.w3.org/2000/01/rdf-schema#"
>
    <rdf:Description rdf:about=
        "http://www.videodetective.com/titledetails.aspx?publishedid=390615">
        <_3:film.film.performance.actor rdf:resource=
            "http://www.videodetective.com/actordetails.aspx?performerid=4998"/>
        <_3:film.film.performance.actor rdf:resource=
            "http://www.videodetective.com/actordetails.aspx?performerid=7349"/>
        <_3:film.film.performance.actor rdf:resource=
            "http://www.videodetective.com/actordetails.aspx?performerid=59108"/>
        <_3:film.film.performance.actor rdf:resource=
            "http://www.videodetective.com/actordetails.aspx?performerid=30903"/>
        <_3:film.film.performance.actor rdf:resource=
            "http://www.videodetective.com/actordetails.aspx?performerid=61052"/>
        <_4:title>BODY OF LIES</_4:title>
    </rdf:Description>
    <rdf:Description rdf:about=
        "http://www.videodetective.com/actordetails.aspx?performerid=4998">
        <rdfs:seeAlso rdf:resource="Leonardo_DiCaprio_4998.xml"/>
    </rdf:Description>
    <rdf:Description rdf:about=
        "http://www.videodetective.com/actordetails.aspx?performerid=30903">
        <rdfs:seeAlso rdf:resource="Oscar_Isaac_30903.xml"/>
    </rdf:Description>
    ... etc.
```

You can now upload the serialized data to a web server and contribute it to the cloud of Linked Data. While making the raw RDF files available is useful, to properly participate in the Linked Open Data community your web server should be configured to handle requests as depicted in Figure 5-2. That is, it should direct people from the URI representing the real-world entity to an information resource that describes it in context. Fortunately, there are well-vetted recipes available online for configuring the

Apache server to do this. One place to start is the W3C's "Best Practices Recipes for Publishing RDF Vocabularies" at *http://www.w3.org/TR/swbp-vocab-pub/*.

Once you have uploaded all the files to a server, you have successfully created and published a Linked Data set! Someone else could now publish another file that lists which of your URLs are the "sameAs" movie URLs from Wikipedia, Netflix, IMDB, or a movie review site like Rotten Tomatoes. Then a crawler (similar to the one you built in Chapter 5) could construct queries that worked across all of these datasets. The goal of Linked Open Data is to answer questions like, "Which movies starring Kevin Bacon that were favored by more than 80% of critics are available for rent right now?" without any one person or company owning all of the data.

This example generates static RDF/XML files about movies. This data isn't constantly changing, and therefore it can be useful even if the static files are updated only occasionally. However, for applications in which the data *is* constantly changing, it makes more sense to dynamically generate the RDF/XML on the fly.

For instance, if we had access to a live message board server, we could extend `message_board_to_sioc.py` to generate live Linked Data using a simple WSGI server. WSGI (Web Server Gateway Interface) is a standard method for interfacing Python applications and web servers. In Chapter 10 we will use CherryPy, a simple Python application server, but for our current purposes we will use the SimpleServer built into the WSGI reference implementation, which is now a standard Python library.

The message board data has a number of interesting facets that systems might want information about: users, subjects, and messages. We will limit ourselves to dealing with users and messages, but you should be thinking about how you could extend it (with just a few extra lines of code) to handle requests for subjects.

Each user will have a unique URI of the form `http://ourserver/users/<userid>`, and similarly we will assign each message a URI of the form `http://ourserver/messages/<messageid>`. When a system attempts to dereference a URI, we will inspect the URI requested to determine whether it is asking for a real-world entity or an information resource. If the request is for a real-world entity, we will redirect the system to the location of the information resource. If the request is for an information resource, we will inspect the request to determine whether it is asking for information about a user or a message, and create an appropriate CONSTRUCT SPARQL query to provide the information requested. The results of the SPARQL query will be serialized into RDF/XML and sent back as the response.

When the server is first initialized, it uses the `message_board_to_sioc` module to load the message board data into the graph that is used by the SPARQL queries. If you had access to a live relational database, you could modify the code so that instead of dumping the complete SQL database into a graph at startup, you could populate a small in-memory graph on each request that contained only the resources necessary to fulfill the inquiry specified by the URI.

The code for the complete server is available at *http://semprog.com/psw/chapter7/mes sage_board_LOD_server.py.* Place the server file in the same directory as the message_board_to_sioc module and SQL data, and execute the file:

```
import os
from wsgiref import simple_server

import rdflib
from rdflib import Namespace

import message_board_to_sioc
from message_board_to_sioc import SIOC, DC, DCTERMS, FOAF, RDFS, MB

"""
A very simple Linked Open Data server
It does not implement content negotiation (among other things)
...and only serves rdf+xml
"""

server_addr =  "127.0.0.1"
server_port = 8000
infores_uri_component = "/rdf"

def rewrite(environ):
    #add infores first path segment
    return "http://" + environ["HTTP_HOST"] + infores_uri_component + \
        environ["PATH_INFO"]

def test_handler(environ):
    resp = {"status":"200 OK"}
    resp["headers"] = [("Content-type", "text/html")]
    outstr = ""
    for k in environ.keys():
        outstr += str(k) + ": " + str(environ[k]) + "<br>"
    resp["body"] = [outstr]
    return resp

def redirect(environ):
    resp = {"status":"303 See Other"}
    resp["headers"] = [("Location", rewrite(environ))]
    return resp

def servedata(environ):
    #Additional ns' for the queries
    ourserver = "http://" + server_addr + ":" + str(server_port) + "/"
    MBMSG = Namespace(ourserver + "messages/")
    MBUSR = Namespace(ourserver + "users/")

    path = environ["PATH_INFO"]

    resp = {"status":"200 OK"}
    resp["headers"] = [("Content-type", "application/rdf+xml")]

    if environ["PATH_INFO"].find("users") != -1:
        #user request query
```

```python
            userid = "mbusr:" + path[path.rindex("/") + 1:]
            query = """CONSTRUCT {
                            """ + userid + """ sioc:creator_of ?msg .
                    ?msg dc:title ?title .
                            """ + userid + """ foaf:name ?name .
                } WHERE {
                    ?msg sioc:has_creator """ + userid + """ .
                        ?msg dc:title ?title .
                            """ + userid + """ foaf:name ?name .
                    } """
        else:
                #message request query
            msgid = "mbmsg:" + path[path.rindex("/") + 1:]
            query = """CONSTRUCT {
                            """ + msgid + """ dc:title ?title .
                            """ + msgid + """ sioc:has_creator ?user .
                            """ + msgid + """ sioc:content ?content .
                        } WHERE {
                            """ + msgid + """ dc:title ?title .
                            """ + msgid + """ sioc:has_creator ?user .
                            """ + msgid + """ sioc:content ?content .
                } """

        bindingdict = {'sioc':SIOC,
                        'dc':DC,
                        'dcterms':DCTERMS,
                        'foaf':FOAF,
                        'rdfs':RDFS,
                        'mb':MB,
                    'mbmsg':MBMSG,
                    'mbusr':MBUSR}

        resp["body"] = [sg.query(query, initNs=bindingdict).serialize(format='xml')]

        return resp

def error(environ, errormsg):
    resp = {"status":"400 Error"}
    resp["headers"] = [("Content-type", "text/plain")]
    resp["body"] = [errormsg]

def application(environ, start_response):
    """Dispatch based on first path component"""
    path = environ["PATH_INFO"]

    if path.startswith(infores_uri_component):
        resp = servedata(environ)

    elif path.startswith("/messages/") or \
        path.startswith("/users/"):

            resp = redirect(environ)

    elif path.startswith("/test/"):
        resp = test_handler(environ)
```

```
        else:
            resp = error(environ, "Path not supported")

        start_response(resp["status"], resp["headers"])
        if resp.has_key("body"):
            return resp["body"]
        else:
            return ""


if __name__ == "__main__":

    #initialize the graph
    if not os.path.exists('message_board.db'):
        message_board_to_sioc.load_data('message_board.sql', 'message_board.db')

    serverlocation = server_addr + ":" + str(server_port)

    #change the MB namespace to the base URI for this server
    message_board_to_sioc.MB = Namespace("http://" + serverlocation + "/")
    sg = message_board_to_sioc.message_board_to_sioc('message_board.db')

    httpd = simple_server.WSGIServer((server_addr, \
        server_port),simple_server.WSGIRequestHandler)
    httpd.set_app(application)
    print "Serving on: " + serverlocation + "..."
    httpd.serve_forever()
```

As a test, try accessing the test_handler method, which will dump out the contents of
the WSGI request environment dictionary, by pointing your web browser to *http://127
.0.0.1:8000/test/*. If everything is working, you should see something like this:

```
PATH_INFO: /test/
SERVER_PORT: 8000
HTTP_KEEP_ALIVE: 300
HTTP_ACCEPT_CHARSET: ISO-8859-1,utf-8;q=0.7,*;q=0.7
REMOTE_HOST: localhost
HTTP_ACCEPT_ENCODING: gzip,deflate
...
```

Now try fetching the data for user 5 using the URI http://127.0.0.1:8000/users/5.
Depending on how your web browser is configured to handle content with the mime
type application/rdf+xml, you may either be prompted to save a file or you'll see the
raw RDF/XML in your browser frame. If your system has wget or curl, you can also
test the server by making requests from the command line using something like this:

```
$ wget -q -O - http://127.0.0.1:8000/users/5
```

 If you get an error about the CONSTRUCT query, your RDFLib may
not be current. Users have reported problems with CONSTRUCT queries on various platform configurations prior to 2.4.1. To obtain the latest
RDFLib from SVN, see the Tip on page 93.

Viewing raw XML and using wget are rather unsatisfying ways of browsing Linked Data. And while Linked Data is published primarily for the benefit of applications, it is extremely useful to get a taste of the data your application will be consuming by inspecting the data yourself. Happily, a number of applications allow you to peruse the cloud of data.

Linked Data browsers come in two forms: browser extensions and hosted browsers. Both approaches have advantages and drawbacks, so it is useful to know about the various offerings. Linked Data browsers typically try to connect the various graphs you are traversing by building a local graph of data you have encountered. This is useful, as it lets you see how various data sources augment one another, but depending on the sources that you access, the cache of data can get both very large and out-of-date if the sources are changing rapidly (as is common in a development environment).

Browser extensions are very useful if the Linked Data you are working with is not available on the public Internet (such as when you are working on a private test server). Browser-based Linked Data viewers are generally limited by the size of the local graph they can maintain. Hosted browsers often have greater computational resources for managing a graph cache, but they are limited to data sources that are publicly available. Here is a short list of Linked Data browsers you may want to utilize:

Tabulator

Tabulator is a Firefox plug-in that allows you to look at Linked Data from a number of different perspectives. The ability to easily switch between serialized RDF views and trees with expandable branches allows you to explore the links and is a useful way to debug Linked Data output.

More information can be found at *http://dig.csail.mit.edu/2007/tab/*.

Disco

Disco is a hosted "hyperdata" browser that, like Tabulator, automatically traverses `owl:sameAs` and `owl:seeAlso` links and retrieves other resources referenced by the data.

More information can be found at *http://www4.wiwiss.fu-berlin.de/rdf_browser/*.

OpenLink Data Explorer

The OpenLink Data Explorer (ODE) extension is a Firefox plug-in that is something of a hybrid Linked Data browser. It orchestrates a number of different (hosted) Linked Data services to provide a view into various types of data.

More information can be found at *http://ode.openlinksw.com/*.

The WSGI server we created is rather bare-bones, but it does fulfill the requirements for serving Linked Data. Still, there are some obvious modifications you may want to make. For one thing, the server makes no attempt to actually "negotiate" the type of data it provides—ideally, it should inspect the `Accept` header provided by the client and, based on the server's preferences, return an accommodating form of data. For instance, if the client indicated that it preferred receiving RDF in N3 format, the server could modify the `format` parameter in the serialization call for the graph produced by the CONSTRUCT SPARQL query. Alternatively, if the client doesn't know how to consume RDF and has a preference for HTML, it would be good to return a redirect to a URI that produces human-friendly output.

If you do make modifications to the server, you can verify that its behavior still conforms to the Linked Data community's best practices by using the Vapour test suite available on SourceForge (*http://vapour.sourceforge.net/*). If your modifications to the server are available on the public Internet, you can use a hosted version of Vapour available from *http://validator.linkeddata.org/vapour*.

## PART III

Putting It into Practice

## CHAPTER 8

Overview of Toolkits

By now you have absorbed plenty of theory and practice, you understand the basics of semantic web standards, and you're totally convinced that you need to build your next big application using semantics. Right? So now it's time to learn about industrial-strength toolkits that you can use to power a web- or client-server application.

A number of good graph-store solutions have emerged since developers first started building semantic web technologies. In this chapter we'll provide an overview of several open source and commercial offerings, but we'll primarily be looking at Sesame, which is one of the leading graph stores and is considered to have excellent performance. We'll first look at the Sesame APIs, then we'll work through the complete process of installing a Java Web Server (if you don't already have Jetty or Tomcat set up), installing Sesame, using the workbench, and adding data. We've also provided a small Python module that allows you to query Sesame directly from Python, and you'll see how to create a simple application that queries and updates the Sesame store.

Finally, we'll look at some of the widgets from the MIT SIMILE project, such as Exhibit and Timeline, which are designed to make the exploration and visualization of semantic data easy. You'll learn how to hook the widgets directly to the Sesame server and build web applications that allow users to search and visualize data in the graph without any server-side programming.

Sesame

Sesame is an open source Java framework for querying and storing RDF data; it was originally developed by the Dutch company Aduna as a research prototype for the European Union research project On-To-Knowledge. It's currently developed as community project and hosted at *http://openrdf.org*. We've chosen to use Sesame for the examples in this book primarily because it has an excellent administration interface included in the distribution, because it's easy to install, and because of its strong performance.

You'll need a Java Runtime Environment (JRE) that works with at least Java 5 to use Sesame. Mac OS X and many Linux distributions come with Java pre-installed. If you don't already have it, you can download the latest version from *http://java.sun.com/javase/downloads/*.

Like RDFLib, Sesame can be embedded in applications, but unlike RDFLib, it can also be used in a standalone server mode, much like a traditional database with multiple applications connecting to it. Before we use it as a server, we will briefly look at how to use Sesame through its native API in order to learn about some of its features.

Using the Sesame Java API

Sesame uses a modular architecture that allows capabilities to be snapped together as needed. The native Java APIs gives you great control over using as many or as few of these capabilities as you desire. To get started, we will create a class that wraps a Sesame repository and encapsulates the most frequently used graph operations, allowing us to use Sesame in much the same way we have been using RDFLib.

You can build this class as we walk through the example, or you can download the completed class from *http://semprog.com/psw/chapter8/SimpleGraph.java*. Although the SimpleGraph class is useful on its own, it's really just a starting point from which to explore various capabilities of Sesame for your own needs.

We will start by creating the class and adding two constructors. The first constructor instantiates a bare-bones in-memory graph repository; the second will take a boolean, which, when true, creates an in-memory store that supports a form of forward-chaining inference directly in the data store. Like our previous encounters with forward-chaining inference, this one incorporates a set of rules that automatically expands certain RDFS information in your ontologies into additional assertions. So, for example, classes that are declared as rdfs:subClassOf of a parent class will have rdf:type information corresponding to both their class and their parent class asserted in the graph:

```
import org.openrdf.query.*;
import org.openrdf.model.vocabulary.*;
import org.openrdf.repository.*;
import org.openrdf.repository.sail.SailRepository;
import org.openrdf.sail.inferencer.fc.ForwardChainingRDFSInferencer;
import org.openrdf.sail.memory.MemoryStore;
import org.openrdf.rio.*;
import org.openrdf.model.*;

import java.net.URL;
import java.net.URLConnection;
import java.util.*;
import java.io.*;

public class SimpleGraph {

    Repository therepository = null;

    // useful -local- constants
    static RDFFormat NTRIPLES = RDFFormat.NTRIPLES;
```

```
static RDFFormat N3 = RDFFormat.N3;
static RDFFormat RDFXML = RDFFormat.RDFXML;
static String RDFTYPE =  RDF.TYPE.toString();

/**
 *  In memory Sesame repository without inferencing
 */
public SimpleGraph(){
    this(false);
}

/**
 * In memory Sesame repository with optional inferencing
 * @param inferencing
 */
public SimpleGraph(boolean inferencing){
    try {
        if (inferencing){
        therepository =
            new SailRepository(new ForwardChainingRDFSInferencer(new
                MemoryStore()));

        } else {
            therepository = new SailRepository(new MemoryStore());
        }
        therepository.initialize();
    } catch (RepositoryException e) {
        e.printStackTrace();
    }
}
}
```

The SimpleGraph class provides access to a select set of Sesame's high-level (or repository-layer) APIs. The repository layer is itself a high-level abstraction, shielding us from the details of Sesame's lower-level Storage And Inference Layer (Sail). The low-level Sail components manage the actual persistence and manipulation of data and are configurable, allowing us to mix and match various storage options through a set of "stackable" components. When we pass true to the SimpleGraph constructor, it stacks the memory-based graph store under the RDFS inferencing component and returns this configuration wrapped in an easy-to-use Sail repository interface.

We could further configure the MemoryStore to save its state to disk on a periodic basis, allowing us to recover the state of the graph on the next instantiation of the class. To do this, we need only pass a file path (string) to the MemoryStore constructor. Alternatively, we could replace the MemoryStore altogether with an interface-compatible NativeStore, which would allow us to efficiently expand the size of the graph beyond the limitations of main memory. To make this change, replace the call to the MemoryStore constructor with a call to the NativeStore constructor (org.openrdf.sail.nativerdf.NativeStore), along with a java.io.File object that will be used to store the disk image of the graph, e.g., new NativeStore(new File("/file/path")).

Next, we will add some methods for creating objects like URIrefs, BNodes, and literals. As you see from the rest of the methods in our class, the repository's `getConnection()` method provides access to the repository for high-level data manipulation. Once a repository connection has been obtained, it should be released whether the operation was successful or not; hence, we wrap all of the connections in try/finally blocks. Obviously, a good implementation of `SimpleGraph` should do something more useful on exceptions than simply printing a stack trace:

```
/**
 *  Literal factory
 *
 * @param s the literal value
 * @param typeuri uri representing the type (generally xsd)
 * @return
 */
public org.openrdf.model.Literal Literal(String s, URI typeuri) {
    try {
        RepositoryConnection con = therepository.getConnection();
        try {
            ValueFactory vf = con.getValueFactory();
            if (typeuri == null) {
                return vf.createLiteral(s);
            } else {
                return vf.createLiteral(s, typeuri);
            }
        } finally {
            con.close();
        }
    } catch (Exception e) {
        e.printStackTrace();
        return null;
    }
}

/**
 * Untyped Literal factory
 *
 * @param s the literal
 * @return
 */
public org.openrdf.model.Literal Literal(String s) {
    return Literal(s, null);
}

/**
 *  URIref factory
 *
 * @param uri
 * @return
 */
public URI URIref(String uri) {
    try {
        RepositoryConnection con = therepository.getConnection();
        try {
```

```
                ValueFactory vf = con.getValueFactory();
                return vf.createURI(uri);
            } finally {
                con.close();
            }
        } catch (Exception e) {
            e.printStackTrace();
            return null;
        }
    }

    /**
     * BNode factory
     *
     * @return
     */
    public BNode bnode() {
        try{
            RepositoryConnection con = therepository.getConnection();
            try {
                ValueFactory vf = con.getValueFactory();
                return vf.createBNode();
            } finally {
                con.close();
            }
        }catch(Exception e){
            e.printStackTrace();
            return null;
        }
    }
```

With the factories in place to create the raw pieces of an RDF statement, let's add
methods to populate the graph. The first method allows us to assert raw triples into
the graph composed of the objects obtained from the factory methods we just added.

The addString, addFile, and addURI methods take serialized RDF and add it to the
repository. The format parameter is used to select the proper RDF parser for the serialization being loaded (and in the case of addURI, to set the proper HTTP ACCEPT
header for content negotiation). These parsers are a part of Sesame's modular architecture and are managed through the RIO (RDF I/O) package. RIO components can
be used outside of Sesame for handling RDF serialization, and they can be augmented
as new standards emerge without affecting the core Sesame systems:

```
    /**
     * Insert Triple/Statement into graph
     *
     * @param s subject uriref
     * @param p predicate uriref
     * @param o value object (URIref or Literal)
     */
    public void add(URI s, URI p, Value o) {
        try {
            RepositoryConnection con = therepository.getConnection();
            try {
```

```
                    ValueFactory myFactory = con.getValueFactory();
                    Statement st = myFactory.createStatement((Resource)
                        s, p, (Value) o);
                    con.add(st);
            } finally {
                con.close();
            }
        }
        catch (Exception e) {
            // handle exception
        }
}

/**
 *  Import RDF data from a string
 *
 * @param rdfstring string with RDF data
 * @param format RDF format of the string (used to select parser)
 */
public void addString(String rdfstring,  RDFFormat format) {
    try {
        RepositoryConnection con = therepository.getConnection();
        try {
            StringReader sr = new StringReader(rdfstring);
            con.add(sr, "", format);
        } finally {
            con.close();
        }
    } catch (Exception e) {
        e.printStackTrace();
    }
}

/**
 *  Import RDF data from a file
 *
 * @param location of file (/path/file) with RDF data
 * @param format RDF format of the string (used to select parser)
 */
public void addFile(String filepath,  RDFFormat format) {
    try {
        RepositoryConnection con = therepository.getConnection();
        try {
            con.add(new File(filepath), "", format);
        } finally {
            con.close();
        }
    } catch (Exception e) {
        e.printStackTrace();
    }
}

/**
 *  Import data from URI source
 *  Request is made with proper HTTP ACCEPT header
```

```
 *   and will follow redirects for proper LOD source negotiation
 *
 * @param urlstring absolute URI of the data source
 * @param format RDF format to request/parse from data source
 */
public void addURI(String urlstring, RDFFormat format) {
    try {
        RepositoryConnection con = therepository.getConnection();
        try {
            URL url = new URL(urlstring);
            URLConnection uricon = (URLConnection) url.openConnection();
            uricon.addRequestProperty("accept", format.getDefaultMIMEType());
            InputStream instream = uricon.getInputStream();
            con.add(instream, urlstring, format);
        } finally {
            con.close();
        }
    } catch (Exception e) {
        e.printStackTrace();
    }
}
```

Now that we can get things into the graph, let's add methods for getting information out of the repository. The first method, dumpRDF, is simply an RDF serialization of everything that's in the repository. The tuplePattern method, like RDFLib's Graph.triples() method, allows us to search for specific patterns of triples in the graph, using null to specify wildcards. Finally, we'll add two methods for running SPARQL queries on the graph. The first runSPARQL method can be used for SPARQL queries of the CONSTRUCT or DESCRIBE form, which return a new graph construction where the format parameter tells the system how the new graph should be returned. The other runSPARQL method can be used for queries of the SELECT form, which return a Java List of solutions and bindings:

```
/**
 *   dump RDF graph
 *
 * @param out output stream for the serialization
 * @param outform the RDF serialization format for the dump
 * @return
 */
public void dumpRDF(OutputStream out, RDFFormat outform) {
    try {
        RepositoryConnection con = therepository.getConnection();
        try {
            RDFWriter w = Rio.createWriter(outform, out);
            con.export(w);
        } finally {
            con.close();
        }
    } catch (Exception e) {
        e.printStackTrace();
    }
}
```

```java
/**
 * Convenience URI import for RDF/XML sources
 *
 * @param urlstring absolute URI of the data source
 */
public void addURI(String urlstring) {
    addURI(urlstring, RDFFormat.RDFXML);
}


/**
 * Tuple pattern query - find all statements with the pattern, where null
 * is a wildcard
 *
 * @param s subject (null for wildcard)
 * @param p predicate (null for wildcard)
 * @param o object (null for wildcard)
 * @return serialized graph of results
 */
public List tuplePattern(URI s, URI p, Value o) {
    try{
        RepositoryConnection con = therepository.getConnection();
        try {
            RepositoryResult repres = con.getStatements(s, p, o, true);
            ArrayList reslist = new ArrayList();
            while (repres.hasNext()) {
                reslist.add(repres.next());
            }
            return reslist;
        } finally {
            con.close();
        }
    }catch(Exception e){
        e.printStackTrace();
    }
    return null;
}

/**
 * Execute a CONSTRUCT/DESCRIBE SPARQL query against the graph
 *
 * @param qs CONSTRUCT or DESCRIBE SPARQL query
 * @param format the serialization format for the returned graph
 * @return serialized graph of results
 */
public String runSPARQL(String qs, RDFFormat format) {
    try{
        RepositoryConnection con = therepository.getConnection();
        try {
            GraphQuery query =
                con.prepareGraphQuery(
                org.openrdf.query.QueryLanguage.SPARQL, qs);
            StringWriter stringout = new StringWriter();
```

```
                RDFWriter w = Rio.createWriter(format, stringout);
                query.evaluate(w);
                return stringout.toString();
            } finally {
                con.close();
            }
        }catch(Exception e){
            e.printStackTrace();
        }
        return null;
    }


    /**
     * Execute a SELECT SPARQL query against the graph
     *
     * @param qs SELECT SPARQL query
     * @return list of solutions, each containing a hashmap of bindings
     */
    public List runSPARQL(String qs) {
        try{
            RepositoryConnection con = therepository.getConnection();
            try {
                TupleQuery query =
                    con.prepareTupleQuery(
                    org.openrdf.query.QueryLanguage.SPARQL, qs);
                TupleQueryResult qres = query.evaluate();
                ArrayList reslist = new ArrayList();
                while (qres.hasNext()) {
                    BindingSet b = qres.next();
                    Set names = b.getBindingNames();
                    HashMap hm = new HashMap();
                    for (Object n : names) {
                        hm.put((String) n, b.getValue((String) n));
                    }
                    reslist.add(hm);
                }
                return reslist;
            } finally {
                con.close();
            }
        }catch(Exception e){
            e.printStackTrace();
        }
        return null;
    }
}
```

Now we will build a simple test class to exercise the various methods in our
SimpleGraph wrapper class. We will load RDF into the repository through a URI, as raw
triples and from a string, and then we'll ask for the data using a tuple pattern, two
SPARQL queries, and finally a dump of the whole repository.

Create the following class and save it as SimpleTest.java in the same directory as SimpleGraph.java (or download it from *http://semprog.com/psw/chapter8/SimpleTest .java*):

```java
import java.util.List;

import org.openrdf.model.URI;
import org.openrdf.model.Value;

public class SimpleTest {

    public static void main(String[] args) {
        // a test of graph operations
        SimpleGraph g = new SimpleGraph();

        // get LOD from a URI -  Jamie's FOAF profile from Hi5
        g.addURI("http://api.hi5.com/rest/profile/foaf/241087912");

        // manually add a triple/statement with a URIref object
        URI s1 = g.URIref("http://semprog.com/people/toby");
        URI p1 = g.URIref(SimpleGraph.RDFTYPE);
        URI o1 = g.URIref("http://xmlns.com/foaf/0.1/person");
        g.add(s1, p1, o1);

        // manually add with an object literal
        URI s2 = g.URIref("http://semprog.com/people/toby");
        URI p2 = g.URIref("http://xmlns.com/foaf/0.1/nick");
        Value o2 = g.Literal("kiwitobes");
        g.add(s2, p2, o2);

        // parse a string of RDF and add to the graph
        String rdfstring = "<http://semprog.com/people/jamie>
            <http://xmlns.com/foaf/0.1/nick> \"jt\" .";
        g.addString(rdfstring, SimpleGraph.NTRIPLES);

        System.out.println("\n==TUPLE QUERY==\n");
        List rlist = g.tuplePattern(null,
            g.URIref("http://xmlns.com/foaf/0.1/nick"), null);
        System.out.print(rlist.toString());

        // run a SPARQL query - get back solution bindings
        System.out.println("\n==SPARQL SELECT==\n");
        List solutions = g.runSPARQL("SELECT ?who ?nick " +
                "WHERE { " +
                    "?x <http://xmlns.com/foaf/0.1/knows> ?y . " +
                    "?x <http://xmlns.com/foaf/0.1/nick> ?who ." +
                    "?y <http://xmlns.com/foaf/0.1/nick> ?nick ."   +
                "}");
        System.out.println("SPARQL solutions: " + solutions.toString());

        // run a CONSTRUCT SPARQL query
        System.out.println("\n==SPARQL CONSTRUCT==\n");
        String newgraphxml = g.runSPARQL("CONSTRUCT { ?x
            <http://semprog.com/simple#friend> ?nick . } " +
                "WHERE { " +
```

```
            "?x <http://xmlns.com/foaf/0.1/knows> ?y . " +
            "?x <http://xmlns.com/foaf/0.1/nick> ?who ." +
            "?y <http://xmlns.com/foaf/0.1/nick> ?nick ."    +
        "}", SimpleGraph.RDFXML);
    System.out.println("SPARQL solutions: \n" + newgraphxml);

    // dump the graph in the specified format
    System.out.println("\n==GRAPH DUMP==\n");
    g.dumpRDF(System.out, SimpleGraph.NTRIPLES);
    }
}
```

Sesame is conveniently packaged in a number of different forms for different types of deployments. For our `SimpleGraph` class, we will use the One-JAR library distribution (*openrdf-sesame-2.2.4-onejar.jar* at the time of this writing) available from *http://www .openrdf.org/download.jsp*. Download *sesame-onejar.jar* and place it in the same directory as the `SimpleGraph` and `SimpleTest` classes.

To compile these classes from the command line, type:

```
$ javac -cp openrdf-sesame-2.2.4-onejar.jar SimpleGraph.java SimpleTest.java
```

To use the Sesame libraries, you will also need to configure a Java logger. Sesame uses the Simple Logging Facade for Java (SLF4J), which allows you to connect Sesame to your favorite Java logger. The easiest way to get started is to download the latest SLF4J distribution from *http://www.slf4j.org/download.html* and unpack it. In the distribution you will find the two files *slf4j-simple.jar* and *slf4j-api.jar*. Copy these JARs to the same directory as the *sesame-onejar.jar* file and the `SimpleGraph` and `SimpleTest` classes.

To run the test class from the command line, type:

```
$ java -cp
    openrdf-sesame-2.2.4-onejar.jar:slf4j-api-1.5.6.jar:slf4j-simple-1.5.6.jar:.
    SimpleTest
```

Note that you may need to adjust the classpath (`-cp`) file separators as appropriate for your system. On Windows, this would be a semicolon (;) instead of a colon (:).

RDFS Inferencing in Sesame

In Chapter 6 we discussed how you can infer the type of a resource based on the domain and range of properties that reference it. While your application could take on this and other types of inferencing responsibilities, you can delegate many of these responsibilities by using a semantic platform.

Sesame can conduct a wide range of inferencing about RDFS type relations. In this example we will see how Sesame can infer the type of an object based on the `rdfs:subClassOf` relation. This will allow us to avoid writing code like that in Chapter 6, where we had to walk the type hierarchy to see if one class was the parent of another.

Let's start by creating another simple test class called TypeTest. This class will make use of the film ontology you built with Protégé in Chapter 6. The ontology is available at *http://semprog.com/psw/chapter6/film-ontology.owl*. The main method will load the ontology, which declares that both the Actor and Director classes are rdfs:subClassOf of Person.

The ontology file also provides some sample instance data that declares that Harrison_Ford is of rdf:type Actor, whereas Ridley_Scott is of rdf:type Director, but in neither case does it say explicitly that either one is of rdf:type Person. The RDFS inferencer operates on the instance data as it is being loaded into the repository and generates new assertions of rdf:type for each instance's parent type.

Save this class as TypeTest.java (or download it from *http://semprog.com/psw/chapter8/ TypeTest.java*):

```java
import java.util.List;

import org.openrdf.model.URI;
import org.openrdf.model.Value;

public class TypeTest {
    public static void main(String[] args) {

        // create a graph with type inferencing
        SimpleGraph g = new SimpleGraph(true);

        // load the film schema and the example data
        g.addFile("film-ontology.owl", SimpleGraph.RDFXML);

        List solutions = g.runSPARQL("SELECT ?who WHERE  { " +
          "?who <http://www.w3.org/1999/02/22-rdf-syntax-ns#type>
             <http://semprog.com/film#Person> ." +
               "}");
        System.out.println("SPARQL solutions: " + solutions.toString());
    }
}
```

Compile the TypeTest class and run it:

```
$ javac -cp openrdf-sesame-2.2.4-onejar.jar SimpleGraph.java TypeTest.java
$ java -cp
    openrdf-sesame-2.2.4-onejar.jar:slf4j-api-1.5.6.jar:slf4j-simple-1.5.6.jar:.
    TypeTest
```

If everything goes as planned, we should learn that both Harrison Ford and Ridley Scott are people.

Behavior-Oriented Programming with Elmo

We have said that semantic programming is really about producing actions consistent with a model given some data. Elmo is an extension of Sesame that allows you to focus on this goal without being distracted by the tasks of managing RDF data. By encapsulating the behavior of ontologies, Elmo allows you to write programs at the modeling level, rather than at the RDF triple level.

Elmo uses Java annotations to facilitate the use of patterns such as composition, separation of concerns, and aspect-oriented approaches for mapping behaviors onto models. The Elmo distribution also contains a code generator for turning RDFS and OWL ontology files into Java classes that can be used to drive the Elmo applications. (In the next chapter we'll develop a similar but less complete system in Python that does not use code generation.)

In this example we make use of the ontology class files that come in the Elmo distribution for well-known ontologies such as FOAF. The example creates an Elmo manager that handles connections to the RDF, fetches Tim Berners-Lee's FOAF information, and then iterates through all resources in the file that are of rdf:type foaf:Person. It then uses the Java abstraction of the FOAF ontology to access and print the foaf:names of the people in his file:

```
import java.net.URL;
import org.openrdf.concepts.foaf.Person;
import org.openrdf.elmo.*;
import org.openrdf.elmo.sesame.*;
import org.openrdf.rio.RDFFormat;

public class ElmoDemo {

 public static void main(String[] args) {
  ElmoModule module = new ElmoModule();
  SesameManagerFactory factory = new SesameManagerFactory(module);
  SesameManager manager = factory.createElmoManager();

  try {
   URL url = new URL("http://www.w3.org/People/Berners-Lee/card.rdf");
   manager.getConnection().add(url, null, RDFFormat.RDFXML);
  } catch (Exception e) {
   e.printStackTrace();
  }
  for (Person person : manager.findAll(Person.class)) {
   System.out.print("Name: ");
   System.out.println(person.getFoafNames());
  }
 }
}
```

To compile this example, you will need the base *elmo* JAR, the *elmo-codegen* and *elmo-foaf* JARs from the Elmo distribution, the *javaassist* and *persistence-api* JARs from the *lib* directory that comes with the Elmo distribution, and the *sesame-onejar* that we used previously. (The command lines have been broken up across multiple lines for readability.)

```
javac -cp elmo-1.4.jar:elmo-codegen-1.4.jar:elmo-foaf-1.4.jar:javaassist-3.7.ga.jar:
    persistence-api-1.0.jar:openrdf-sesame-2.2.4-onejar.jar:. ElmoDemo.java
```

Running this will list the `foaf:name`(s) of all the people of `rdf:type foaf:Person` (based on the `Person.class`):

```
java -cp elmo-1.4.jar:elmo-codegen-1.4.jar:elmo-foaf-1.4.jar:javaassist-3.7.ga.jar:
    persistence-api-1.0.jar:openrdf-sesame-2.2.4-onejar.jar:. ElmoDemo
```

A Servlet Container for the Sesame Server

Now that we have seen how Sesame can be embedded in an application, let's set up Sesame as a standalone server. This will give us additional flexibility as we embark on more sophisticated projects, and will allow us to access Sesame using other programming languages.

You can skip this section if you already have a Java Servlet container like Tomcat or Jetty that supports the Java Servlet API 2.4 and JSP 2.0 or newer. If you don't have one yet, we recommend using Jetty, as it has a very simple installation and will get you up and running quickly.

The official site for Jetty is *http://jetty.mortbay.org/*. You can download the latest distribution at *http://dist.codehaus.org/jetty/*. At the time of writing, the latest version is 6.1.14, but you're probably safe just downloading the latest release (not prerelease) version. The download will be a ZIP file, which you should unzip wherever you want to run Jetty—there is no installation procedure.

Now, go to the directory where you unzipped Jetty and type `java -jar start.jar`. You should see something like this:

```
$ java -jar start.jar
2009-01-05 21:21:55.346::INFO:  Logging to STDERR via org.mortbay.log.StdErrLog
2009-01-05 21:21:55.488::INFO:  jetty-6.1.14
...
2009-01-05 21:21:59.636::INFO:  Opened
    /Users/bbts/writing/jetty-install/jetty-6.1.14/logs/2009_01_06.request.log
2009-01-05 21:21:59.651::INFO:  Started SelectChannelConnector@0.0.0.0:8080
```

Visit *http://localhost:8080* in your web browser and you should see a welcome page confirming that you successfully installed Jetty. Go back to your prompt and hit Ctrl-C to stop it.

Installing the Sesame Web Application

The Sesame servlets are also very simple to install. The download page for Sesame is *http://www.openrdf.org/download.jsp*. Download the archive of the latest version (this chapter uses version 2.2.3, *openrdf-sesame-2.2.3-sdk.zip*) and extract it.

There should be a directory within the archive called *war*. This directory contains two files, *openrdf-sesame.war* and *openrdf-workbench.war*, which are web archives for your

Java server. Simply copy these files into the *webapps* directory of your Jetty (or Tomcat) installation. When you restart Jetty, you should see a message telling you that these two WAR files were detected and installed as new applications.

The Workbench

As mentioned earlier, one of the coolest things about Sesame is that it comes with a very functional administration interface. You can access this interface by visiting *http://localhost:8080/openrdf-workbench* in your web browser, which should look something like Figure 8-1.

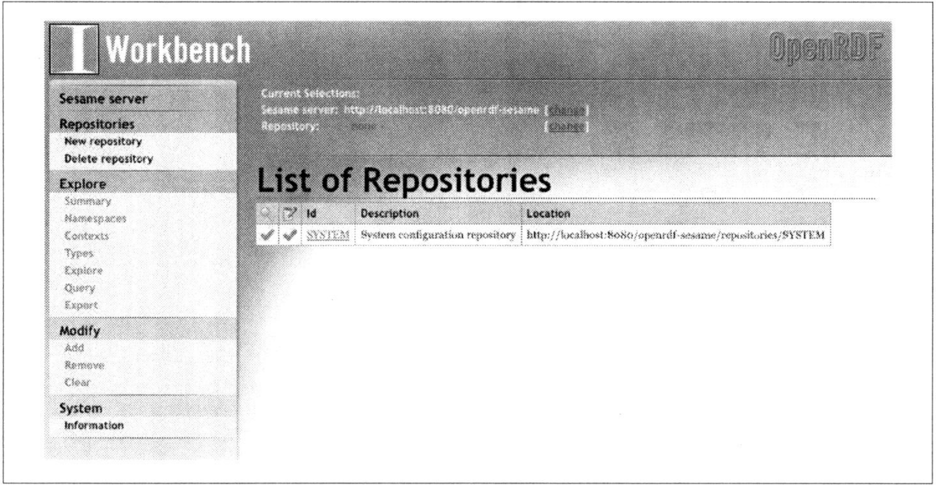

Figure 8-1. The OpenRDF Sesame workbench

The menu items on the left side allow you to access administration pages to create new data repositories (graphs), add RDF data, then explore and query the data. This section will walk you through setting up a repository and filling it with the movie data from Chapter 7. To start, click "New Repository" and you'll see a form like Figure 8-2.

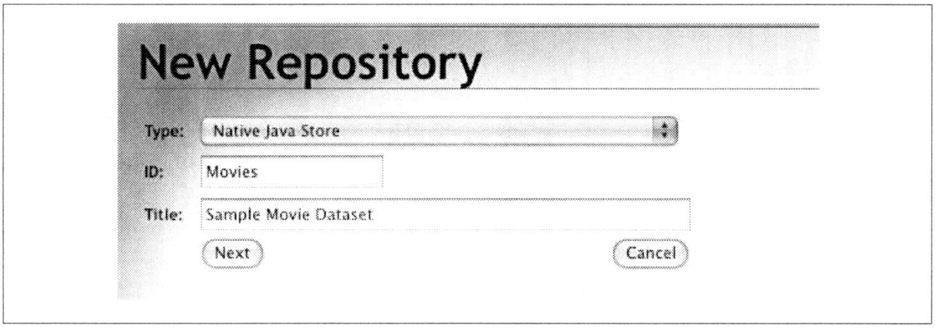

Figure 8-2. Creating a movie repository

There are three fields to fill in here: the ID, which is just a short name for the repository and the one you'll refer to later when accessing the repository through the API; the title, which is a longer description; and the type, which is the storage mechanism for this repository. There are nine options for the type:

1. In-Memory Store
2. In-Memory Store RDF Schema
3. In-Memory Store RDF Schema and Direct Type Hierarchy
4. Native Java Store
5. Native Java Store RDF Schema
6. Native Java Store RDF Schema and Direct Type Hierarchy
7. MySQL RDF Store
8. PostgreSQL RDF Store
9. Remote RDF Store

Options 1–3 are different kinds of In-Memory Store. This is the fastest type, since the entire graph is kept in memory. Persistence (keeping the graph between restarts of the server) is optional but works well—the graph is saved to disk frequently and loaded completely into memory when the Sesame server is started. The main drawback of this is that you need to guarantee that the graph will never grow larger than the memory available to the server. This may seem like a severe restriction, but remember that graphs of hundreds of thousands of triples can easily fit in memory in modern machines, so this is feasible for many applications.

The difference between options 1, 2, and 3 is whether they support the various types of ontology inferencing that you learned about in Chapter 6. Since more sophisticated inferencing uses more resources, you're given the choice of how much you need. Option 2 provides for RDF Schema inferencing, which can infer object types that aren't directly specified using properties, and option 3 adds type hierarchy inferencing on top of that, which can infer even more about types by using a specified class hierarchy.

Options 4–6 are similar to 1–3, but they store all the data on disk in a Sesame-specific format. This obviously reduces performance, but it removes the restriction that the entire graph must fit in memory. In benchmarks, this type usually performs better than the MySQL and PostgreSQL options (7 and 8). Again, you have the choice to include inferencing at the level you need it.

Using options 7 and 8, you can store your graph in a currently running MySQL or PostgreSQL server. Although these are usually slightly slower than the native Java store, they have the advantage of using your existing relational databases, which is great if you work at an organization that already has people who back up and maintain these databases. Since the native Sesame store isn't something that most people are familiar with, it's probably lower-risk to go with a relational database if you work with people who already deal with them.

Finally, option 9 lets you point to a repository that is kept on a different Sesame server, and expose it through this server. This is useful if you have one or more instances of your repository on machines that you don't want people connecting to directly, either for security purposes or because you want to be able to change the location of the backend repository without reconfiguring the client applications.

For now, create a native Java store repository with the ID "Movies", as shown earlier in Figure 8-2.

Adding Data

Once you've created the repository, the first thing you'll need to do is get some data into it. Sesame lets you do this by uploading a file, pointing to a URL, or directly typing/pasting RDF statements. Clicking on "Add" in the menu on the left side of the admin interface will reveal an interface similar to the one shown in Figure 8-3.

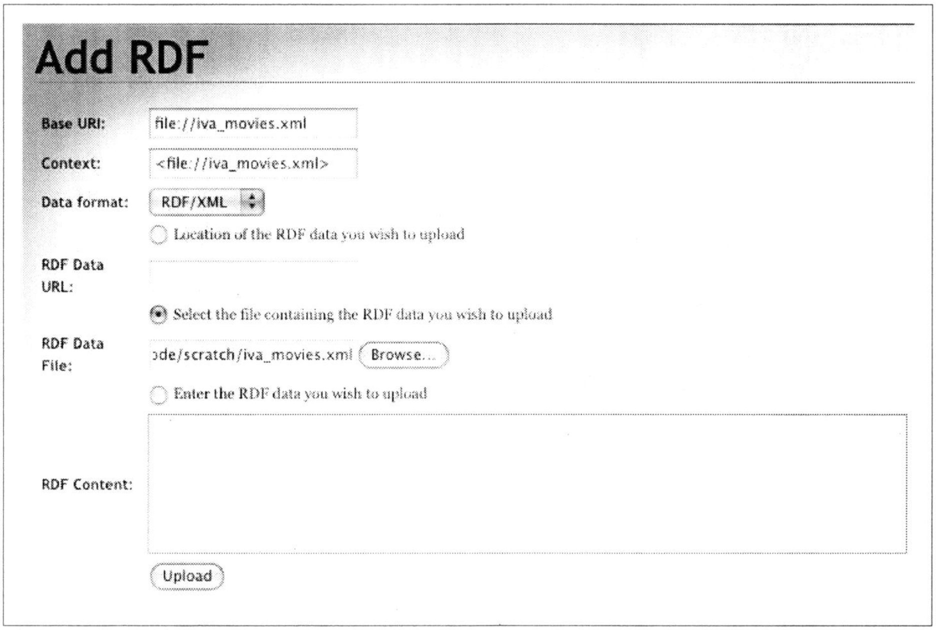

Figure 8-3. Adding data to the movie repository

We're going to add the movie data that we generated in Chapter 7. To get the data, you can download *http://semprog.com/psw/chapter7/iva_movies.xml*, or you can generate the data file from *iva_movies.py* like this:

```
$ python iva_movies.py > iva_movies.xml
```

Make sure the data format is set to "RDF/XML" and click "Browse..." to locate *iva_movies.xml*. After you've selected it, you'll notice that the Base URI and Context fields automatically get filled in. *Base URI* is the namespace that will be used for nodes in your file that don't have a namespace specified (generally, and in this case, there aren't any nodes without specified namespaces).

Context is something we haven't come across before—it is stored along with all the triples in the file so that you know where they come from. If the same triple is added with a different context (perhaps you upload a different file), there will be two copies with separate contexts. This allows you to remove all the triples from one file if, for example, you determine that the data in it is bad, without removing identical triples that came from sources that are still considered good. This is another clear win over traditional data modeling, which usually doesn't record the origin of links between tables, much less allow multiple links with different origins.

Click on "Upload" to add your data to the repository.

SPARQL Queries

The Sesame workbench also allows you to query the repository using a web form. This is excellent for checking the data, answering ad hoc questions, and testing queries for use in your applications. Clicking on "Query" in the left menu bar will bring up a page like the one shown in Figure 8-4.

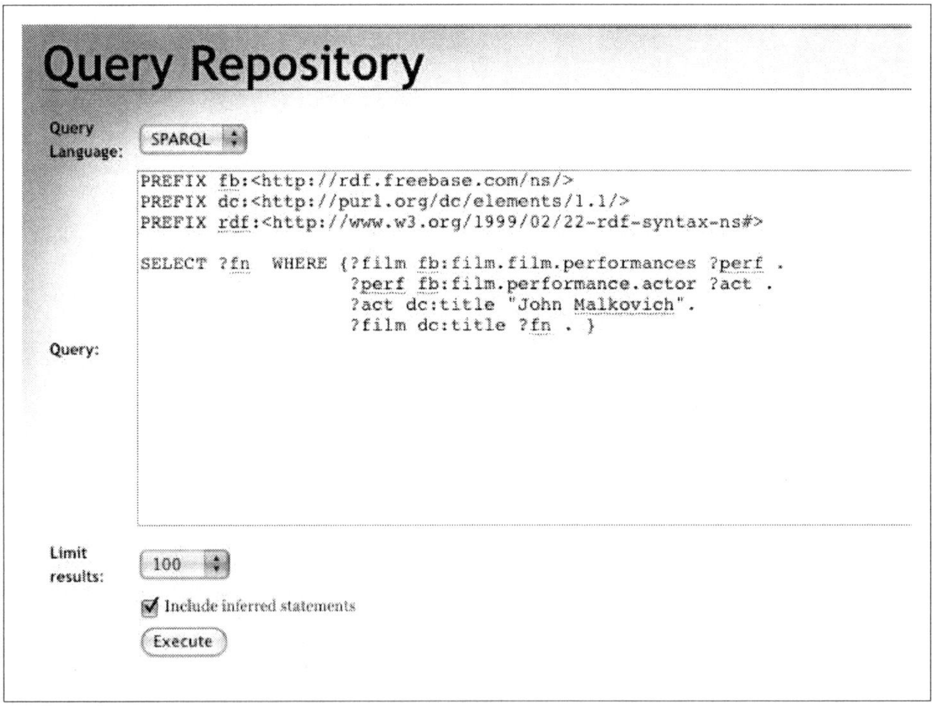

Query Repository

Query Language:	SPARQL

```
PREFIX fb:<http://rdf.freebase.com/ns/>
PREFIX dc:<http://purl.org/dc/elements/1.1/>
PREFIX rdf:<http://www.w3.org/1999/02/22-rdf-syntax-ns#>

SELECT ?fn  WHERE {?film fb:film.film.performances ?perf .
                   ?perf  fb:film.performance.actor ?act .
                   ?act dc:title "John Malkovich".
                   ?film dc:title ?fn . }
```

Query:

Limit results: 100

☑ Include inferred statements

(Execute)

Figure 8-4. Querying the movie store

When you uploaded the RDF file, it included some namespace prefixes. These have now become some of the repository's default prefixes. When the query form appears, the prefixes are automatically included in the main text box, which means you can type your SPARQL query without having to define all the namespaces every time. In Figure 8-4, there's a query for all of the films that have performances by an actor named "John Malkovich". If you enter this query and click "Execute," you'll be taken to a new page with the results, shown in Figure 8-5. SPARQL SELECT query results are formatted into very clean tables with all the requested fields as headers. The fields themselves are all links that you can click on to see everything in the repository around a specific node.

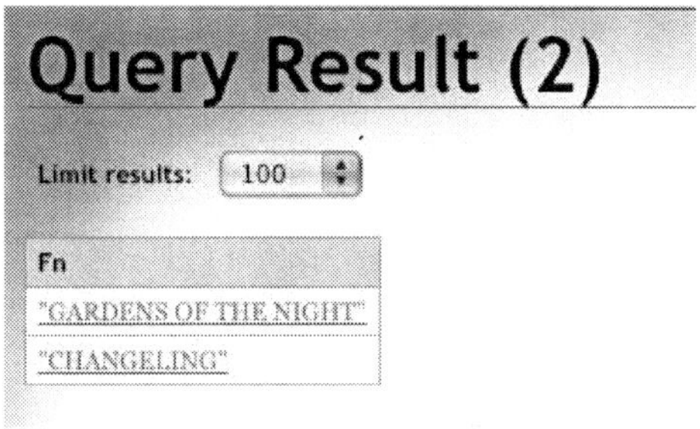

Figure 8-5. Results of the query

If you like, you can try the following more complex query, which shows all the people who have costarred in a film with John Malkovich:

```
PREFIX fb:<http://rdf.freebase.com/ns/>
PREFIX dc:<http://purl.org/dc/elements/1.1/>
PREFIX rdf:<http://www.w3.org/1999/02/22-rdf-syntax-ns#>

SELECT ?costar ?fn WHERE {?film fb:film.film.performances ?p1 .
                ?film dc:title ?fn .
                ?p1 fb:film.performance.actor ?a1 .
                ?a1 dc:title "John Malkovich".
                ?film fb:film.film.performances ?p2 .
                ?p2 fb:film.performance.actor ?a2 .
                ?a2 dc:title ?costar .}
```

We recommend that you become familiar with the other features of the workbench. You can explore your data by node type, namespace, context, or through queries. You can add data by typing N3 notation directly, and you can remove triples by specifying any combination of subject, predicate, object, or context. For example, you could decide that IVA messed up its John Malkovich data and remove anything that came from the IVA context about John Malkovich.

REST API

To actually use Sesame in an application, you'll want to be able to access it from other applications. This is usually done through the Sesame REST API, which is documented at *http://www.openrdf.org/doc/sesame2/system/ch08.html*. Like many other REST APIs, calls are made by passing parameters in a URL through a regular HTTP request, and the server returns the result in a machine-readable format. Sesame returns its results in JSON, which is great because that's very easy to parse in Python.

For example, you could turn the previous query into a REST request:

```
http://<server>/openrdf-sesame/repositories/celebs?
    query=select+?fn+where+%7B?film+fb:film.film.performances+?p1+.+?film ...etc...
```

and get a set of JSON with variable bindings:

```
{"headers": ["costar", "fn"],
 "data": [{"costar": {"type": "literal", "value": "Angelina Jolie"},
           "fn": {"type": "literal", "value": "CHANGELING"},
          {"costar": {"type": "literal", "value": "Gillian Jacobs"},
           "fn": {"type": "literal", "value": "GARDENS OF THE NIGHT"},
    ...
```

To make this easy to use from Python, we've created a simple module called *pysesame.py* that wraps the REST API. You can download this module from *http://semprog.com/psw/chapter8/pysesame.py*. The code should be easy enough to translate into other languages if you're not planning your project in Python. Here's what it looks like:

```python
from urllib import urlopen,quote_plus
from simplejson import loads

class connection:
    def __init__(self,url):
        self.baseurl = url
        self.sparql_prefix = ""

    def addnamespace(self, id, ns):
        self.sparql_prefix += 'PREFIX %s:<%s>\n' % (id,ns)

    def __getsparql__(self, method):
        data = urlopen(self.baseurl + method).read()
        try:
            result = loads(data)['results']['bindings']
            return result
        except:
            return [{'error':data}];

    def repositories(self):
        return self.__getsparql__('repositories')

    def use_repository(self, r):
        self.repository = r
```

```
def query(self, q):
    q = 'repositories/' + self.repository + '?query=' +
        quote_plus(self.sparql_prefix + q)
    return self.__getsparql__(q)
```

This module defines a class called `connection`, which represents a connection to a Sesame store. In reality, however, there is no persistent connection—it's just a wrapper to store settings for making the REST requests. The `connection` class is initialized with the name of the server. After initialization, you can call `repositories` to see a list of repositories on that server, `use_repository` to choose one, `addnamespace` to define namespaces, and finally `query` to query the database using SPARQL.

The module also includes an example in its main method so you can see it in action:

```
if __name__ == '__main__':
    c = connection('http://localhost:8080/openrdf-sesame/')
    c.use_repository('Movies')
    c.addnamespace('fb','http://rdf.freebase.com/ns/')
    c.addnamespace('dc','http://purl.org/dc/elements/1.1/')
    res = c.query("""SELECT ?costar ?fn WHERE {?film fb:film.film.performances ?p1 .
                    ?film dc:title ?fn .
                    ?p1 fb:film.performance.actor ?a1 .
                    ?a1 dc:title "John Malkovich".
                    ?film fb:film.film.performances ?p2 .
                    ?p2 fb:film.performance.actor ?a2 .
                    ?a2 dc:title ?costar .}""")
    for r in res: print r
```

Running *pysesame.py* from the command line will show you the results of this sample query:

```
$ python pysesame.py
{u'costar': {u'type': u'literal', u'value': u'Gillian Jacobs'},
 u'fn': {u'type': u'literal', u'value': u'GARDENS OF THE NIGHT'}}
{u'costar': {u'type': u'literal', u'value': u'Ryan Simpkins'},
 u'fn': {u'type': u'literal', u'value': u'GARDENS OF THE NIGHT'}}
{u'costar': {u'type': u'literal', u'value': u'John Malkovich'},
 u'fn': {u'type': u'literal', u'value': u'GARDENS OF THE NIGHT'}}
```

If you want to try some other queries, you can either modify the main method in *pysesame* or, more practically, create a new Python file and just add the imports at the beginning, like this:

```
from pysesame import connection, use_repository, addnamespace, query
```

We'll be using *pysesame* a lot more in Chapter 10 when we build a real application. For now, just make sure you're familiar with the workbench, and try importing some other RDF files and querying them through the REST API.

Other RDF Stores

We chose to give detailed coverage of Sesame because it's powerful, easy to set up, and easy to work with, but there are many other excellent ways to store semantic data.

There are many abandoned projects in this space, so we think it's important to make sure you know which ones are still being actively maintained. We can't cover all the different RDF storage options in as much detail as Sesame, but here are some brief descriptions of other options you may encounter.

We encourage you to explore these options on your own, or at least remember their names so you're not caught off guard when they come up at your next cocktail party.

Jena (Open Source)

Jena is another Java-based open source semantic web toolkit. It came out of Hewlett-Packard's Semantic Web Research Lab and is similar to Sesame in the functionality that it provides. Jena's APIs are much more complete, however; it has built-in support for OWL (and other ontology languages), several different rule-based reasoners, SPARQL query support, and several persistent graph stores. Because of its research origin, Jena has a wide variety of new and experimental libraries and contributions. The downside to all this is that the learning curve for Jena tends to be significantly steeper than for Sesame. You can download Jena from *http://jena.sourceforge.net/*.

Redland (Open Source)

Redland is a mature set of open source C libraries that work in concert to deliver a complete semantic platform. The libraries are modular, providing RDF parsing, query capability, and triple storage as separable elements that can be integrated into other projects. The Redland project also provides library bindings for Perl, PHP, Python, and Ruby. In addition to the libraries, Redland comes with a handy command-line tool called rapper that allows you to fetch, parse, and reserialize RDF to and from a wide variety of formats including N-Triples, N3, Turtle, and RDFa. Rapper also provides an option for generating Graphviz dot-file output to create RDF visualizations like those we created in Chapter 3. You can learn more about the Redland libraries at *http://librdf .org*.

Mulgara (Open Source)

Mulgara is a Java-based open source triplestore. It is a fork from Kowari, an earlier open source project currently owned by Northrop Grumman. Mulgara's functionality is similar to Sesame, with SPARQL query support. Mulgara has support for a large number of J2EE APIs, including the JTA API that allows for transaction support. You can download Mulgara from *http://www.mulgara.org/*.

OpenLink Virtuoso (Commercial and Open Source)

Virtuoso is a highly scalable, object-relational database system that supports standard relational/SQL operations as well as RDF/SPARQL operations. OpenLink software is very involved in the Linked Open Data community, providing the server system used for DBpedia (the data source queried in the Linked Data application in Chapter 5) and the OpenLink Data Extension for Firefox described in Chapter 7. You can learn about Virtuoso and download both commercial and open source distributions from *http:// virtuoso.openlinksw.com*.

Franz AllegroGraph (Commercial)

Franz Inc. is probably best known for its commercial Lisp tools such as Allegro Common Lisp, but it's also been developing semantic web tools for some time and is beginning to feature products like AllegroGraph more prominently on its website. AllegroGraph is a commercial triplestore, compatible with the Sesame API and usable from Java, Python, Lisp, and other languages. You can download a free version of AllegroGraph that will work with up to 50 million triples from *http://agraph.franz.com/*.

Oracle (Commercial)

Oracle Enterprise 11g supports an optional Spatial module that includes native support for RDF, including inference capabilities for RDFS and OWL. RDF data can be queried using SQL, and by using the SEM_MATCH keyword you can include SPARQL-like graph patterns in your queries. You can learn more about Oracle's semantic technology offerings at *http://www.oracle.com/technology/tech/semantic_technologies*.

SIMILE/Exhibit

SIMILE (Semantic Interoperability of Metadata and Information in unLike Environments) is a set of open source projects conducted by the MIT Computer Science and AI Lab and MIT Libraries. Its goal is to simplify the reuse and sharing of data, and to build tools that make it easy to work with data. One of the most successful projects is called Exhibit, which allows you to very easily create rich interactive web pages by including a small amount of JavaScript (much like you would with Google Maps). These pages display semantically structured data. Figure 8-6 shows a few examples of pages that can be built with Exhibit, such as maps, timelines, and galleries, all with very detailed client-side filtering. You can see more examples at *http://simile.mit.edu/exhibit*.

Exhibit includes a lot of different ways to display data. In this section we'll teach you the basics of Exhibit, including how to get it to read RDF, hook it up to Sesame directly, and build a couple of example views. This should be enough to get you started building rich frontends to your semantic applications.

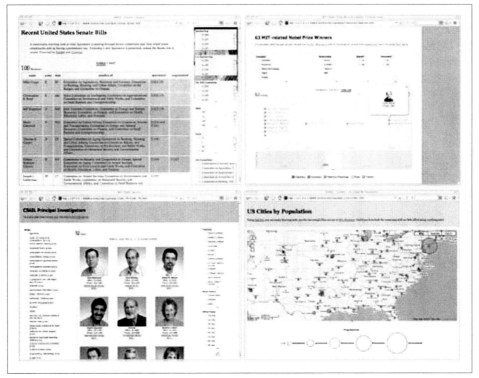

Figure 8-6. Examples of displaying data with Exhibit

A Simple Exhibit Page

The first thing we'll do is use Exhibit to display some RDF data. We've turned our celebrity relationship data from Chapter 2 into an RDF file, which you can view or download at *http://semprog.com/psw/chapter8/celebs.rdf*. Here's a bit of what it looks like:

```
<rdf:Description rdf:about="http://rdf.freebase.com/ns/en/woody_allen">
  <rdf:type rdf:resource="http://rdf.freebase.com/ns/celebrities.celebrity"/>
  <dc:title>Woody Allen</dc:title>
</rdf:Description>
<rdf:Description rdf:about=
    "http://rdf.freebase.com/ns//guid/9202a8c04000641f8000000007c92468">
  <rdf:type rdf:resource=
    "http://rdf.freebase.com/ns/celebrities.romantic_relationship"/>
  <fb:with rdf:resource="http://rdf.freebase.com/ns//en/mia_farrow"/>
  <fb:with rdf:resource="http://rdf.freebase.com/ns//en/woody_allen"/>
  <fb:end>1992</fb:end>
  <fb:start>1980</fb:start>
</rdf:Description>
```

If you look at the file, you'll see that there are basically two types of objects: people, which have the type `celebrities.celebrity` along with their names, and relationships, which have references to the people involved as well as start and end dates.

Now point your browser to *http://semprog.com/psw/chapter8/celebs_basic_exhibit .html*. Wait a moment while the RDF file loads, and then you'll see all the objects in the file displayed as a list. Even this very basic view is interactive. You have the option to change the sort order of the data, which also changes the grouping. Figure 8-7, for example, shows what happens when you reverse sort by the "with" field. You can also click on the linked entities (e.g., "woody_allen" or "mia_farrow") to go to their entries.

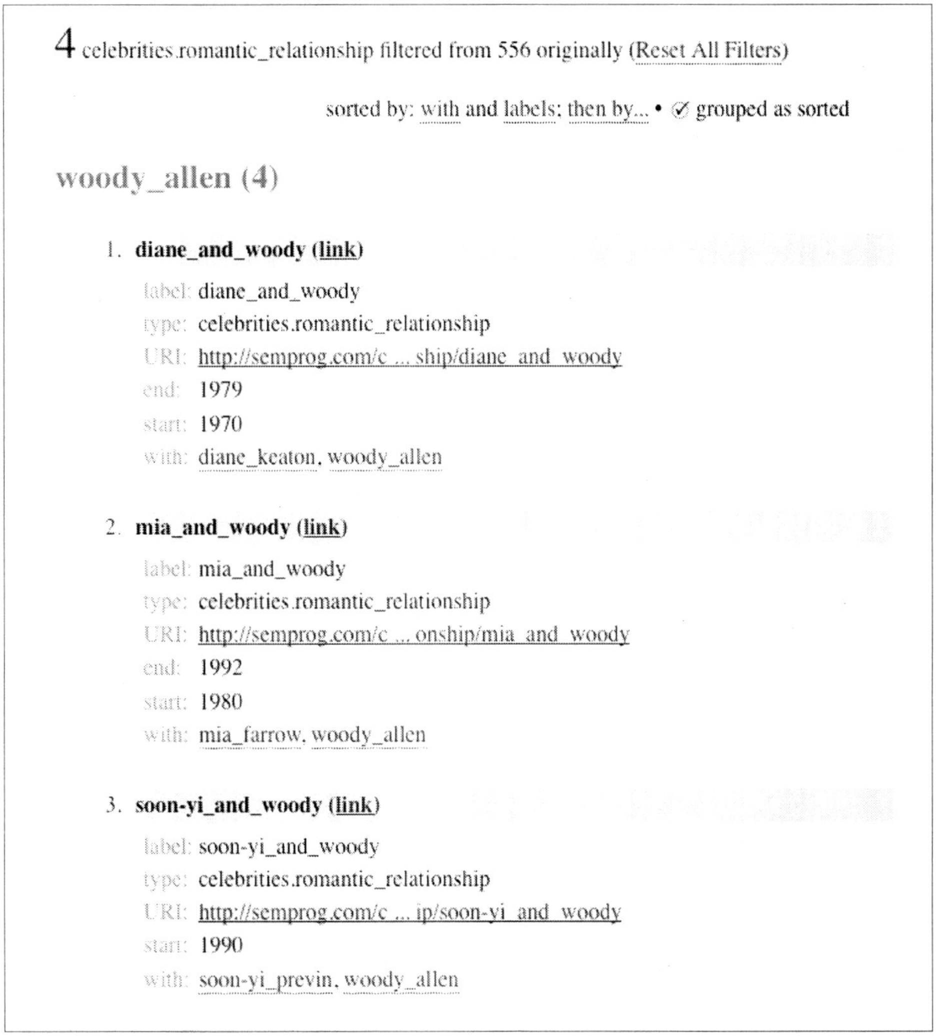

4 celebrities.romantic_relationship filtered from 556 originally (Reset All Filters)

sorted by: with and labels; then by... • ☑ grouped as sorted

woody_allen (4)

1. **diane_and_woody (link)**
   label: diane_and_woody
   type: celebrities.romantic_relationship
   URI: http://semprog.com/c ... ship/diane_and_woody
   end: 1979
   start: 1970
   with: diane_keaton, woody_allen

2. **mia_and_woody (link)**
   label: mia_and_woody
   type: celebrities.romantic_relationship
   URI: http://semprog.com/c ... onship/mia_and_woody
   end: 1992
   start: 1980
   with: mia_farrow, woody_allen

3. **soon-yi_and_woody (link)**
   label: soon-yi_and_woody
   type: celebrities.romantic_relationship
   URI: http://semprog.com/c ... ip/soon-yi_and_woody
   start: 1990
   with: soon-yi_previn, woody_allen

Figure 8-7. Basic celebrity view in Exhibit, reverse-sorted by relationship-with

Obviously this is not a particularly attractive way to view the data, but it's the starting point for an Exhibit application to which you can add more features. Looking at the HTML source of this file (either download the file or "view source" in your browser), you can see that the first part of the file looks like this:

```
<html>
<head>
    <title>Celebrities and Relationships</title>
    <link rel="exhibit/data" type="application/rdf+xml"
        href="celebs.rdf" />
    <script src="http://static.simile.mit.edu/exhibit/api-2.0/exhibit-api.js"
            type="text/javascript"></script>
</head>
```

This is the familiar HTML head section, and we've added a couple of important tags here. One of these is a script include for the Exhibit API, which is being loaded directly from its home on the MIT website. This script exposes the basic Exhibit functionality for loading and displaying semantic data; as you'll see later, we'll need to include additional scripts to get some of the fancier widgets onto the page.

Also necessary is the link tag, which tells Exhibit where to find the data file. Since link tags can be used for many things, such as CSS or RSS files, we set the rel attribute of the link tag to exhibit/data. Exhibit also supports many import formats—in this case we're using RDF, so we include type="application/rdf+xml" in the tag. The advantage of using a link tag rather than just setting a JavaScript variable is that files specified in a link tag can reside on another server, so you could build an application on your own server that displays someone else's RDF file, or that even pulls together data from many different servers and displays it on a single page!

The remainder of the HTML sets up the very basic Exhibit page:

```
<body>
    <h1>Celebrities and Relationships</h1>
    <table width="100%">
        <tr valign="top">
            <td ex:role="viewPanel">
                <div ex:role="view"></div>
            </td>
            <td width="25%">
                browsing controls here...
            </td>
        </tr>
    </table>
</body>
</html>
```

The only things you need to notice here are the tags with ex:role attributes. The line <div ex:role="view"> simply tells Exhibit to set up the default view of key/value pairs for every property in the file.

Searching, Filtering, and Prettier Views

Like many of the things we like, Exhibit makes it very easy to get started but is also very customizable and extensible. To see what it might look like when you add a lot of custom filters, direct your browser to *http://semprog.com/psw/chapter8/celeb_fancy_exhibit.html*. In addition to the sort options you had before, you should now have options for searching and filtering. You'll also notice that the entities themselves are displayed very differently. Take a look at the source code for this more sophisticated page:

```
<table width="100%">
    <tr valign="top">
        <td ex:role="viewPanel" class="mainlist">
            <div ex:role="view" ex:orders=".start">
                <div ex:role="lens" class="lens" style="display: none;">
                    <div class="dateline">
                    From <span ex:content=".start" class="year"></span>
                    to <span ex:content=".end" class="year"></span>:
                    </div>
                    <div class="people" ex:content=".with"></div>
                </div>
            </div>
        </td>
        <td width="25%">
            <div ex:role="facet" ex:facetClass="TextSearch"></div>
            <div ex:role="facet"
                ex:expression=".with" ex:facetLabel="Person"></div>
            <div ex:role="facet" ex:expression=".start"
                ex:facetLabel="Relationship Start"></div>
            <div ex:role="facet" ex:expression=".end"
                ex:facetLabel="Relationship End"></div>
        </td>
    </tr>
</table>
```

There's a lot of new stuff in here, and a lot more you can do with Exhibit. Let's quickly cover the newly introduced tags.

The simplest is `<div ex:role="facet" ex:facetClass="TextSearch">`. This just sets up a basic free text search within the data, which you can see on the top righthand side of the page. You'll also notice that we've changed the "lens" `div` to include an additional property `ex:orders=".with"`, which sets the default search to the `with` property (the people in the relationship) rather than the label, which is just an ugly GUID right now. Sorting is still available; this just affects how it looks when you first load the page.

In the second table cell, we've added some tags that look like this, called *facets*:

```
<div ex:role="facet" ex:expression=".with" ex:facetLabel="Person"></div>
<div ex:role="facet" ex:expression=".start" ex:facetLabel="Relationship Start"></div>
```

Figure 8-8 shows that these add some filtering options to the right side of the page. At the top is a list of people who appear in the `with` field of different relationships, and below are start and end years of these relationships. Beside each entry is the number of records in which it appears. Since we have the most data about Winona Ryder, click on her name and see what happens. The list of objects should filter to only include relationships containing Winona Ryder, but also the **start** and **end** facets filter to only contain the start and end years of Winona's relationships. This is very similar to online shopping sites that let you filter the list down by variables, and filter the other variables so that only the most relevant ones remain.

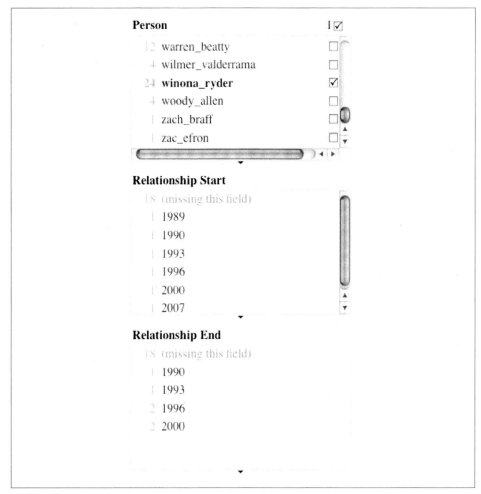

Figure 8-8. Faceted browsing with Winona Ryder selected

Using this faceted search, you could, for example, try to find the longest-lasting celebrity relationship in the dataset that began in 2002.

The final change to the page was creating a lens for displaying the items. Rather than use the default view, which simply shows us every key/value pair, we included a small **div** with the attribute **ex:role="lens"**. This tells Exhibit that it should use the contents of this `div` to display the objects instead of its default. Inside, you'll see several `span` tags, each of which has an attribute `ex:content`. This might look familiar if you've ever used an XML templating language—the `span` is filled in with the contents of the property specified by the `ex:content` attribute. For example, if the `start` attribute of the relationship is 1996, then `<span ex:content=".start"></span>` would become `<span>1996</span>`.

This should be enough information to get you started playing with the basic Exhibit viewer. You can alter the styles, create new lenses, or try creating some more RDF files and building views of them in Exhibit (you can use RDFLib and some of the data files we've provided, or online sources such as FOAF files or Freebase). The only catch is that you won't be able to use a local data source (one that's accessible through the file system or localhost), since the Exhibit server needs to be able to access the file in order to parse it. If you don't have a server to put things on, don't worry—we've provided plenty of RDF for you to play with at *http://semprog.com*.

Linking Up to Sesame

One of the problems with using Exhibit the way we've described is that it has to download the entire data file into your browser before it can be displayed. That's fine with a small file like *celebs.rdf*, but imagine if you had a very large file, say a file of houses for sale across the country. You'd probably want to pre-filter the RDF file to only include one or two cities.

Fortunately, we already have a way to do this. If your data is stored in a Sesame instance, you can use the "construct" form to return an RDF file with a structure that you define, and filter it using the WHERE clause. To see how this might work, consider the simplest example, which just returns everything in the RDF store exactly as it is:

```
CONSTRUCT {?a ?b ?c .} WHERE {?a ?b ?c .}
```

If you go to your Sesame workbench and do this query, you'll just get back a list of triples that exactly match what's in the database. By taking this query and converting it to a REST request, we can dynamically generate an RDF file with the same data. On a Sesame server, the URL looks like this:

```
http://<sesame server>/openrdf-sesame/repositories/celebs?
    query=construct+%7B%3Fx+%3Fy+%3Fz+.%7D+where+%7B%3Fx+%3Fy+%3Fz+.%7D
```

You can easily incorporate this into your Exhibit page by replacing the data link with one that just requests this URL:

```
<link
href="http://<sesame server>/openrdf-sesame/repositories/celebs?
    query=construct+%7B%3Fx+%3Fy+%3Fz+.%7D+where+%7B%3Fx+%3Fy+%3Fz+.%7D"
rel="exhibit/data" type="application/rdf+xml"/>
```

Of course, we haven't really gained anything over using a static file here, since we're dumping out the whole database anyway. However, it's easy to imagine altering the construct query to something like:

```
CONSTRUCT {?a ?b fb:winona_ryder  .} WHERE {?a ?b fb:winona_ryder .}
```

which will only return, in this instance, relationships with Winona Ryder. You could create a set of static pages for each celebrity or city or whatever it is you're interested in, or you could set up the server to dynamically generate the correct link based on user input. We're not going to get into dynamic page generation here, but you'll see more of this in Chapter 10, when we build a whole web application on Sesame and Exhibit.

Timelines

Exhibit supports a lot more than just the object-property view we've covered so far in this chapter. Using various "widgets," it can also show maps, charts, and timelines, which we'll explore in this section. Figure 8-9 shows a timeline in action; you can also see it if you go to *http://semprog.com/psw/chapter8/celeb_timeline.html*. This is a highly interactive widget: you can drag to different years, click in the lower section to jump to a different time period, and click on any of the lines for details of what it represents.

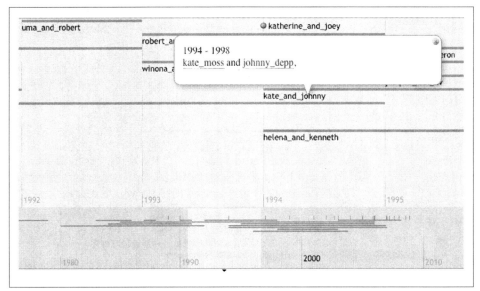

Figure 8-9. Celebrity relationships on a timeline

Looking at the source of the page, you'll notice it's actually quite similar to the previous Exhibit example, with a couple of key differences. The first is that we've included an extra JavaScript file in the header, which is the code for the timeline:

```
<script src=
    "http://static.simile.mit.edu/exhibit/extensions-2.0/time/time-extension.js"
    type="text/javascript"></script>
```

Additional SIMILE widgets usually require additional scripts to be included. The other big change is that we've added some new attributes to the `<div ex:role="view"...>`, which tell Exhibit to show a timeline instead of a regular view:

```
<div ex:role="view"
    ex:viewClass="Timeline"
    ex:start=".start"
    ex:end=".end">
        <div ex:role="lens" class="nobelist-timeline-lens" style="display: none;">
            <div><span ex:content=".start"></span> - <span ex:content=".end">
                </span></div>

            <div>
                <span ex:content=".with"></span>,
            </div>
        </div>
</div>
```

The attribute that turns this into a timeline is simply `ex:viewClass="Timeline"`. After that, there are a couple of properties, `start` and `end`, that are specific to the `Timeline` class. These reference fields in the dataset are interpreted as dates for plotting on the timeline. In the case of the celebrity dataset, we just have years, which are easily interpreted as dates. Again, within the `div` we create another `div` with an attribute `ex:role="lens"`. In this case, however, it describes the pop up that is displayed when you click on any item in the timeline. As you can see from Figure 8-9, we've created a bubble that shows the time range of the relationship and the names of the people involved.

We've only taken you through the basics here—there's much, much more that Exhibit can do. We strongly encourage you to learn more about it by going through some of the tutorials at *http://simile.mit.edu/wiki/Exhibit/For_Authors*. In Chapter 10 you'll see a lot more of Exhibit when we take you through the process of building a whole application using semantic data. But before that, we want to introduce you to some of what we call "semantic programming patterns."

## CHAPTER 9

Introspecting Objects from Data

A common problem you face when working with graph databases is figuring out how to access the data easily and efficiently in your code. By using an OWL ontology to describe the common shape of our data, we can map OWL classes in our graph database to Python objects directly and easily, making it easy to read and write data from the database. This approach to database programming is commonly called *object-relational mapping* and has been popularized in libraries like Python's SQLObject, Java's Hibernate, and Ruby's ActiveRecord. Performing this kind of mapping is very easy with a graph database and an OWL ontology, as we'll demonstrate in this chapter. The RDFObject framework described here demonstrates how easy it is to tightly integrate code with graph data.

RDFObject Examples

Let's get a quick overview of how the RDFObject framework works by looking at some examples using our film data. We've started by loading the OWL ontology into a Sesame repository called `semprog` running on localhost. First we'll initialize the connection to the Sesame repository and create an `RDFObjectFactory`, the base object for creating `RDFObjectGraphs`, which we will use to access our graph data. The `RDFObject Factory` queries the Sesame graph for ontology information on initialization and then pre-computes and caches mappings from Python attributes to RDF URIs:

```
>>> from rdfobject import *
>>> sc = SesameConnection("localhost:8080", "semprog")
>>> factory = RDFObjectGraphFactory(sc)
```

Next, we'll create an `RDFObjectGraph`. The `RDFObjectGraph` caches a small subset of the information in the Sesame graph locally, and it also holds writes and deletes until they are committed. It shouldn't be kept around for a long time, but it should be treated as a per-request object for rendering a web page or updating data from a form:

```
>>> objectGraph = factory.createGraph()
```

Now we'll load an `RDFObject` for the film *Blade Runner*. We'll retrieve the `RDFObjects` for these instances by using RDFLib URIs, but the `RDFObject` wrappers allow us to easily access the properties of these instances. The framework maps Python accessor attributes (such as `name` and `starring`) to corresponding OWL properties (such as `http://www.semprog.com/film#name` and `http://www.semprog.com/film#starring`). Each attribute always returns a list, as there can always be multiple values for a property. If a property points to a URI, the framework always returns another `RDFObject` so that you can chain attributes:

```
>>> filmNs = Namespace("http://www.semprog.com/film#")
>>> bladerunner = objectGraph.get(filmNs["blade_runner"])
>>> print bladerunner.uri
http://www.semprog.com/film#blade_runner
>>> print bladerunner.name[0]
Blade Runner
>>> print bladerunner.type
[<RDFObject http://www.semprog.com/film#Film>]
>>> print bladerunner.starring[0].hasActor[0].name[0]
Harrison Ford
```

Next, we'll load an `RDFObject` for Harrison Ford. You'll notice here that literal values are returned as RDFLib literals and can have a language and datatype associated with them. Also, attributes that are prefixed with `r_` are reversed automatically:

```
>>> harrisonford = objectGraph.get(filmNs["harrison_ford"])
>>> print harrisonford.name
[rdflib.Literal('Harrison Ford', language=None, datatype=...
>>> print harrisonford.r_hasActor[0].r_starring[0].name[0]
Blade Runner
```

Now we'll add the film *Raiders of the Lost Ark* to the graph. To do this, we'll create new `RDFObjects` with appropriate IDs, assign them types, and then link up the relationships. We need to create a `Film` object for the film, a `Role` object for the character Indiana Jones, and a `Performance` object to tie them together. In order for the `RDFObject` to know that an attribute has changes, you need to reset a list into the attribute after you modify the list. At the end, we'll call the `commit` method, which will push all of the changes to the server and flush the local caches as well:

```
>>> raiders = objectGraph.get(filmNs["raiders_of_the_lost_ark"])
>>> raiders.type = [filmNs["Film"]]
>>> raiders.name = ["Raiders of the Lost Ark"]
>>> perf2 = objectGraph.get(filmNs["perf2"])
>>> perf2.type = [filmNs["Performance"]]
>>> indy = objectGraph.get(filmNs["indy"])
>>> indy.type = [filmNs["Role"]]
>>> indy.name = ["Indiana Jones"]
>>> perf2.r_starring = [raiders]
>>> perf2.hasActor = [harrisonford]
>>> perf2.hasRole = [indy]
>>> objectGraph.commit()
```

Now we can access the new data that we've loaded:

```
>>> print indy.name
[rdflib.Literal('Indiana Jones', language=None, datatype=None)]
>>> print raiders.name
```

```
[rdflib.Literal('Raiders of the Lost Ark', language=None, datatype=None)]
>>> print raiders.starring[0].hasActor[0].uri
http://www.semprog.com/film#harrison_ford
```

I can also query for properties by asking for names using the [ ] operator. This lookup doesn't use the OWL ontology and instead looks directly at the links on a URI. This makes it slower, but it's a good way to address data not explicitly described in the ontology. We can't use this approach to do writes, though.

```
>>> print harrisonford["name"][0]
Harrison Ford
>>> print harrisonford["r_hasActor"][0]["r_starring"][0]["name"][0]
Blade Runner
```

This framework is a fast and simple solution for applications like rendering web pages and filling out forms. Now we'll look into how it is built.

RDFObject Framework

You can download the code for the RDFObject framework from *http://semprog.com/psw/chapter9/rdfobject.py*. We'll go through the code here. First, we import libraries and define namespaces that we'll be using:

```
rom rdflib import Namespace, URIRef, Literal, BNode
from rdflib.Graph import Graph
from urllib import quote_plus
from httplib import HTTPConnection
from cStringIO import StringIO
import xml.dom

owlNS = Namespace("http://www.w3.org/2002/07/owl#")
owlClass = owlNS["Class"]
owlObjectProperty = owlNS["ObjectProperty"]
owlDatatypeProperty = owlNS["DatatypeProperty"]
rdfNS = Namespace("http://www.w3.org/1999/02/22-rdf-syntax-ns#")
rdfProperty = rdfNS["Property"]
rdfType = rdfNS["type"]
rdfsNS = Namespace("http://www.w3.org/2000/01/rdf-schema#")
rdfsSubClassOf = rdfsNS["subClassOf"]
rdfsDomain = rdfsNS["domain"]
rdfsRange = rdfsNS["range"]
```

Next, we need to define a class for generating Sesame transaction documents. Sesame uses a simple custom XML over HTTP API for performing updates and deletions, and we'll create a wrapper that builds an XML document from a series of additions and deletions:

```
class SesameTransaction:
    def __init__(self):
        self.trans = xml.dom.getDOMImplementation().\
                    createDocument(None, "transaction", None)

    def add(self, statement):
```

```
        self.__addAction("add", statement)

    def remove(self, statement):
        self.__addAction("remove", statement)

    def toXML(self):
        return self.trans.toxml()

    def __addAction(self, action, statement):
        element = self.trans.createElement(action)
        for item in statement:
            if isinstance(item, Literal):
                literal = self.trans.createElement("literal")
                if item.datatype is not None:
                    literal.setAttribute("datatype", str(item.datatype))
                if item.language is not None:
                    literal.setAttribute("xml:lang", str(item.language))
                literal.appendChild(self.trans.createTextNode(str(item)))
                element.appendChild(literal)
            elif isinstance(item, URIRef):
                uri = self.trans.createElement("uri")
                uri.appendChild(self.trans.createTextNode(str(item)))
                element.appendChild(uri)
            elif isinstance(item, BNode):
                bnode = self.trans.createElement("bnode")
                bnode.appendChild(self.trans.createTextNode(str(item)))
                element.appendChild(bnode)
            else:
                raise Exception("Unknown element: " + item)
        self.trans.childNodes[0].appendChild(element)
```

Now we'll build a connection class for querying and updating a Sesame data store. This class allows SPARQL requests to be made to a Sesame repository via HTTP, with results being returned in RDF XML format and loaded to a local in-memory graph. We'll be making a bunch of CONSTRUCT queries with SPARQL, which is where we'll ask for our result set to come in the form of a graph of triples.

The class also performs updates on the Sesame repository. The API takes in-memory graphs of triples to be added and deleted, turns the two graphs into an XML transaction document, and posts the changes to the server:

```
class SesameConnection:
    def __init__(self, host, repository=None):
        self.host = host
        self.repository = repository
        self.sparql_prefix=""

    def addnamespace(self, id, ns):
        self.sparql_prefix+='PREFIX %s:<%s>\n' % (id,ns)

    def repositories(self):
        return self.__getsparql__('repositories')

    def use_repository(self, r):
```

```
            self.repository = r

    def __request__(self, method, path, data, headers):
        conn = HTTPConnection(self.host)
        conn.request(method, path, data, headers)
        response = conn.getresponse()
        if response.status != 200 and response.status != 204:
            raise Exception("Sesame connection error " +\
                            str(response.status) + " " + response.reason)
        response = response.read()
        conn.close()
        return response

    def query(self, query, graph):
        path = '/openrdf-sesame/repositories/' + self.repository +\
               '?query=' + quote_plus(self.sparql_prefix + query)
        data = self.__request__("GET", path, None, {"accept":"application/rdf+xml"})
        graph.parse(StringIO(data))
        return graph

    def update(self, add=None, remove=None):
        path = '/openrdf-sesame/repositories/' + self.repository + "/statements"
        trans = SesameTransaction()
        if remove is not None:
            for statement in remove: trans.remove(statement)
        if add is not None:
            for statement in add: trans.add(statement)
        data = self.__request__("POST", path, trans.toXML(),
                                {"content-type":"application/x-rdftransaction"})
```

Next, we define a class for storing information about our OWL ontology. The constructor for this class takes a SesameConnection and makes a SPARQL query for all OWL classes and properties. Then the class computes a simplified name for each property, which consists of the last path on the property's URI. For instance, the property http://www.semprog.com/film#directed_by is shortened to directed_by. Reverse properties are also computed, and these are prefixed with r_. These shortened names will be used to refer to the properties as attributes on the RDFObjects:

```
class OWLOntology:
    """
    This class loads the mappings from simple property names
    to OWL property URIs.
    """

    def __init__(self, sesameConnection):
        # query for all OWL classes and properties:
        self._ontGraph = Graph()
        sesameConnection.query(
            """construct {
            ?c rdf:type owl:Class .
            ?c rdfs:subClassOf ?sc .
            ?p rdfs:domain ?c .
            ?p rdfs:range ?d  .
            ?p rdf:type ?pt .
```

```
          } where
          {
           ?c rdf:type owl:Class .
          OPTIONAL {
           ?c rdfs:subClassOf ?sc .
          }
          OPTIONAL {
           ?p rdfs:domain ?c .
           ?p rdfs:range ?d .
           ?p rdf:type ?pt .
          }
          }""", self._ontGraph)
        # map type properties to simplified names:
        self.propertyMaps = {}
        for ontClass in self._ontGraph.subjects(rdfType, owlClass):
            propertyMap = self.propertyMaps[ontClass] = {}
            for property in self._ontGraph.subjects(rdfsDomain, ontClass):
                propertyName = self.getSimplifiedName(property)
                propertyMap[propertyName] = property
            for property in self._ontGraph.subjects(rdfsRange, ontClass):
                propertyName = "r_" + self.getSimplifiedName(property)
                propertyMap[propertyName] = property
        # recursively copy property mappings across the class hierarchy:
        def copySuperclassProperties(ontClass, propertyMap):
            for superclass in self._ontGraph.objects(ontClass, rdfsSubClassOf):
                copySuperclassProperties(superclass, propertyMap)
            propertyMap.update(self.propertyMaps[ontClass])
        for ontClass in self._ontGraph.subjects(rdfType, owlClass):
            copySuperclassProperties(ontClass, self.propertyMaps[ontClass])

    def getSimplifiedName(self, uri):
        if "#" in uri: return uri[uri.rfind("#") + 1:]
        return uri[uri.rfind("/") + 1:]
```

Next, we'll define the RDFObjectGraph class. This class will store the connection and ontology mappings, and it should be instantiated when your application starts up:

```
class RDFObjectGraphFactory:
    """

    A factory for RDFObjects.
    """

    def __init__(self, connection):
        self.connection = connection
        self.connection.addnamespace("xsd", "http://www.w3.org/2001/XMLSchema#")
        self.connection.addnamespace("rdf",
                                "http://www.w3.org/1999/02/22-rdf-syntax-ns#")
        self.connection.addnamespace("rdfs", "http://www.w3.org/2000/01/rdf-schema#")
        self.connection.addnamespace("owl", "http://www.w3.org/2002/07/owl#")
        self.ontology = OWLOntology(connection)

    def createGraph(self):
        return RDFObjectGraph(self.connection, self.ontology)
```

The RDFObjectGraphFactory class is the heart of this framework. The class keeps three in-memory graphs: the first is a cache of triples from the Sesame graph, the second is a graph of triples to be added, and the third is a graph of triples to be removed. The class also holds a cache of instantiated RDFObject instances. The get method checks to see if an RDFObject has already been instantiated; if not, it calls _load to ensure that all adjacent properties for the URI have been loaded:

```
class RDFObjectGraph:
    """
    The RDFObjectGraph caches object values for populating RDFObject values.
    """

    def __init__(self, connection, ontology):
        self._connection = connection
        self._ontology = ontology
        self._rdfObjects = {}
        self._graph = Graph()
        self._added = Graph()
        self._removed = Graph()

    def get(self, uri):
        """
        Gets an RDFObject for the specified URI.
        """
        if uri not in self._rdfObjects:
            self._load(uri)
            self._rdfObjects[uri] = RDFObject(uri, self)
        return self._rdfObjects[uri]

    def _load(self, uri):
        """
        This method ensures that the data for a uri is loaded into
        the local graph.
        """
        if uri not in self._rdfObjects:
            self._connection.query(
            "construct { <" + uri + "> ?p ?o . " +
            "?rs ?rp <" + uri + "> .} where { " +
            "OPTIONAL { <" + uri + "> ?p ?o } " +
            "OPTIONAL { ?rs ?rp <" + uri + "> } }", self._graph)
```

Following are three methods from RDFObjectGraph for retrieving subjects, predicates, and objects. Each method first queries the graph of cached statements from the Sesame server, then removes any statements from the _removed graph, and finally adds statements from the _added graph. This way, changes that are uncommitted still show up when you're querying properties:

```
    def _subjects(self, prop, uri):
        """
        Retrieves all subjects for a property and object URI.
        """
        for triple in self._graph.triples((None, prop, uri)):
            if triple not in self._removed:
                yield triple[0]
        for triple in self._added.triples((None, prop, uri)):
            yield triple[0]
```

```python
def _objects(self, uri, prop):
    """
    Retrieves all objects for a subject URI and property.
    """
    for triple in self._graph.triples((uri, prop, None)):
        if triple not in self._removed:
            yield triple[2]
    for triple in self._added.triples((uri, prop, None)):
        yield triple[2]

def _predicates(self, subject=None, object=None):
    """
    Retrieves all unique predicates for a subject or object URI.
    """
    result = set()
    for triple in self._graph.triples((subject, None, object)):
        if triple not in self._removed:
            result.add(triple[1])
    for triple in self._added.triples((subject, None, object)):
        result.add(triple[1])
    return result
```

Next, we'll add methods for updating subjects and objects, along with a method to commit the changes. Both of the update methods compare the updated list of values against the existing list of values, and add or remove statements as appropriate. Python sets make this an easy operation:

```python
def _setSubjects(self, values, prop, uri):
    """
    Sets all subjects for a property and uri.
    """
    newValues = set(values)
    existingValues = set(self._graph.subjects(prop, uri))
    for value in existingValues - newValues:
        removed = (value, prop, uri)
        self._added.remove(removed)
        self._removed.add(removed)
    for value in newValues - existingValues:
        added = (value, prop, uri)
        self._removed.remove(added)
        self._added.add(added)

def _setObjects(self, uri, prop, values):
    """
    Sets all objects for a uri and property.
    """
    newValues = set(values)
    existingValues = set(self._graph.objects(uri, prop))
    for value in existingValues - newValues:
        removed = (uri, prop, value)
        self._added.remove(removed)
        self._removed.add(removed)
    for value in newValues - existingValues:
        added = (uri, prop, value)
        self._removed.remove(added)
```

```
        self._added.add(added)

    def commit(self):
        """
        Commits changes to the remote graph and flushes local caches.
        """
        self._connection.update(add=self._added, remove=self._removed)
        self._rdfObjects = {}
        self._graph = Graph()
        self._added = Graph()
        self._removed = Graph()
```

Finally, we'll build out the RDFObject class. First we will add a constructor and methods for printing the object:

```
class RDFObject:
    """
    The RDFObject wraps an RDF URI and automatically retrieves property values
    as they are referenced as object attributes.
    """
    def __init__(self, uri, objectGraph):
        self.__dict__["uri"] = uri
        self.__dict__["_objectGraph"] = objectGraph

    def __repr__(self):
        return "<RDFObject " + self.uri + ">"

    def __str__(self):
        return self.uri
```

Next we'll add __getattr__ and __setattr__ methods along with two utility methods. The __getattr__ and __setattr__ methods are special methods called by the Python interpreter when a class attribute is accessed. These methods allow us to intercept the call and dynamically find a value for the named URI, and either return the current values or update the values. Both methods look at the attribute name and structure, and they find the appropriate corresponding property by calling _getProp, which checks the OWL ontology mappings from attributes to URIs:

```
    def __getattr__(self, name):
        self._objectGraph._load(self.uri)
        prop = self._getProp(name)
        if name.startswith("r_"):
            values = self._objectGraph._subjects(prop, self.uri)
        else:
            values = self._objectGraph._objects(self.uri, prop)
        results = self._wrapResults(values)
        return results

    def __setattr__(self, name, values):
        self._objectGraph._load(self.uri)
        unwrappedValues = []
        for value in values:
            # unwrap rdfobjects:
            if isinstance(value, RDFObject):
                unwrappedValues.append(value.uri)
```

```
        # pass through rdflib objects:
        elif isinstance(value, URIRef) or isinstance(value, BNode) or \
            isinstance(value, Literal):
            unwrappedValues.append(value)
        # wrap literals:
        else:
            unwrappedValues.append(Literal(value))
    # look for a property mapping for this name:
    prop = self._getProp(name)
    if name.startswith("r_"):
        self._objectGraph._setSubjects(unwrappedValues, prop, self.uri)
    else:
        self._objectGraph._setObjects(self.uri, prop, unwrappedValues)

def _getProp(self, name):
    if name == "type": return rdfType
    for type in self._objectGraph._objects(self.uri, rdfType):
        propertyMap = self._objectGraph._ontology.propertyMaps[type]
        if name in propertyMap: return propertyMap[name]
    raise AttributeError("Unknown property '" + name + "'")

def _wrapResults(self, results):
    ret = []
    for result in results:
        if isinstance(result, Literal): ret.append(result)
        else: ret.append(self._objectGraph.get(result))
    return ret
```

The __getitem__ method is another special method that implements the functionality of the [ ] operator. This method asks for a list of predicates adjacent to the current URI and tries to find one that matches the attribute name used:

```
def __getitem__(self, key):
    self._objectGraph._load(self.uri)
    # iterate over predicates and look for a matching name:
    reverse = key.startswith("r_")
    if reverse:
        preds = self._objectGraph._predicates(object=self.uri)
        name = key[2:]
    else:
        preds = self._objectGraph._predicates(subject=self.uri)
        name = key
    for pred in preds:
        if self._objectGraph._ontology.getSimplifiedName(pred) == name:
            if reverse:
                values = self._objectGraph._subjects(pred, self.uri)
            else:
                values = self._objectGraph._objects(self.uri, pred)
            return self._wrapResults(values)
    raise KeyError("Property '" + key + "' not found")
```

How RDFObject Works

One of the most difficult aspects of making traditional relational database mapping tools like Hibernate, SQLObject, and ActiveRecord is sorting out the data dependencies between relational database tables and figuring out how those dependencies map to the objects in memory. Because each triple in a graph is independent of all other triples, and triples map pretty much one-to-one to property assignments in objects, it is much easier to map OWL objects to Python classes automatically.

A surprising attribute of graph databases is that they are often easier to extend than the code that lies on top of them, while traditional relational databases often lag behind their application code. This dynamic nature of graph databases opens up a new field of tools and techniques for programming, where the data becomes more central to defining the flow of the application.

The dynamic nature of graph databases is clear in our earlier examples. Additions and changes to the OWL ontology can be immediately realized in the code, and the introspective quality of querying the database for the meaning of the attributes makes prototyping new functionality easy. In the next chapter we'll explore some of the applications that you can build.

## CHAPTER 10

Tying It All Together

We would definitely fall short of our goal of creating a practical guide to using semantic data if we didn't actually tie it all together in the end. The purpose of this chapter is to take many of the ideas, lessons, and technologies that we've covered so far and use them to build a complete application. In doing this, we hope to put into practice the patterns we showed you in the previous chapter, and also to really make clear the advantages of using semantics to represent your data.

Much of the commercial activity (i.e., where people are making money) in semantic web technologies is in building flexible content management systems that work within large companies or that allow for easy data partnerships between groups or companies. Among other things, this chapter should serve as a recipe for building such a system, which we'll do by building an example application.

A Job Listing Application

The focus of our example application is a job listing site for American companies, which will tie together many different kinds of data to show how semantic data makes extensibility and sophisticated searches much easier. Hopefully you'll be able to use this as a guide to building any application based on semantic data.

There are many steps to building this application:

1. Loading an initial dataset into Sesame
2. Setting up a web application server
3. Creating HTML templates for viewing objects
4. Expanding the dataset from public sources
5. Republishing the data for other semantic applications
6. Querying across many datasets
7. Doing visualizations using Exhibit

All these steps will be covered in a tutorial style. We have chosen to use CherryPy, a lightweight but full-featured Python-based application server, on account of its simplicity. Many web frameworks include ORM mappings and tools for working with relational databases, which we don't need here.

Application Requirements

Let's consider a list of requirements for our application. These are typical for all applications, and they really direct us toward using semantic data:

- The application needs to be up and running with very basic functionality in a short time.
- It needs to be flexible to new data as we expand the functionality.
- It should be designed from the beginning to share and consume data using standards.
- Rapid iteration and experimentation should be expected.

Although we've designed these requirements to illustrate our use of semantic data, you'll almost certainly recognize them as needs that appear over and over again. In particular, they arise when designing web applications that will change with user needs, or when embarking on a large enterprise project that may change direction once you determine how it will actually be used.

Job Listing Data

Let's suppose that at the beginning of our project, all we have is a tabular dataset with a list of job postings. The data is divided into columns by employer, title, experience required, etc. We've provided a data file at *http://semprog.com/psw/chapter10/joblist .csv* that looks like this:

```
"Job Title","Salary","Location","Company","Crunchbase","Ticker"
"Tugboat Captain",10000,"San Francisco, CA","Metaweb","metawebtechnologies",
"Ancient Mariner",20000,"New York, NY","Google","google","GOOG"
"Fisherman",30000,"Seattle, WA","Microsoft",,"MSFT"
"Third Mate",40000,"Sunnyvale, CA","Yahoo",,"YHOO"
"Stevedore",50000,"New York, NY","Tumblr","tumblr",
```

Obviously the jobs themselves are made up, but the companies are real (you'll see why later). Every company has a name; for a few of the publicly traded companies, stock ticker symbols are provided, and for some of the startups, references to a well-known startup database called Crunchbase are provided. Later we'll be using these to get more data about the companies, but for now let's concentrate on getting the job listings into a format where we can use them.

Converting to RDF

In order to use this file with a semantic data store, we need to make its semantics explicit and convert it to a standard file format like RDF or N3. When deciding how to describe the semantics, it's always a good idea to look and see if someone has already come up with an RDF schema that suits your dataset, particularly if it's something that's commonly expressed. Job openings are posted on the Web with some frequency, so it's a safe bet that someone has come up with a schema for them. And sure enough, there's a description of a job posting schema at *http://www.medev.ac.uk/interoperability/rss/1.0/modules/jobs/*. Figure 10-1 shows a graph of what the schema looks like.

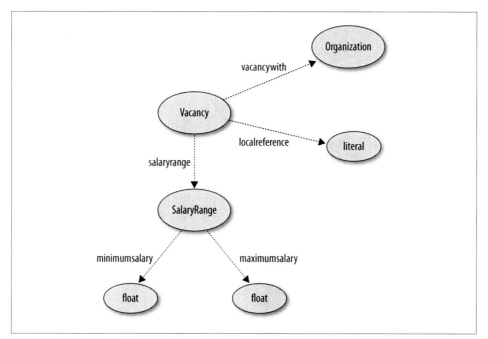

Figure 10-1. Subset of schema for http://www.medev.ac.uk/interoperability/rss/1.0/modules/jobs/

One easy way to do the conversion is to write a small Python script that uses RDFLib. In fact, we've written it for you, and you can download it from *http://semprog.com/psw/chapter10/convert_jobs.py*. Here's what it looks like:

```
from rdflib.Graph import ConjunctiveGraph
from rdflib import Namespace, BNode, Literal, RDF, URIRef
import csv

JOBS = Namespace("http://www.medev.ac.uk/interoperability/rss/"+\
+"1.0/modules/jobs/rss1.0jobsmodule#")
DC = Namespace("http://purl.org/dc/elements/1.1/")
JB = Namespace("http://semprog.com/schemas/jobboard#")
COMPANY = Namespace("http://purl.org/rss/1.0/modules/company/")
RDFS=Namespace('"http://www.w3.org/2000/01/rdf-schema#')

jg = ConjunctiveGraph()
jg.bind('jobs', JOBS)
jg.bind('dc', DC)
```

```
jg.bind('jobboard', JB)
jg.bind('company', COMPANY)
jg.bind('rdfs', RDFS)

# Incremental counter for vacancy IDs
vid = 0

for title, salary, location, company, crunchbase, \
ticker in csv.reader(file('joblist.csv')):
    # Create the vacancy
    vid += 1
    vacancy = JB[str(vid)]
    jg.add((vacancy, RDF.type, JOBS['Vacancy']))
    jg.add((vacancy, DC['title'], Literal(title)))

    location_id = location.lower().replace(' ', '_').replace(',', '')
    jg.add((vacancy, JB['location'], JB[location_id]))
    jg.add((JB[location_id], DC['title'], Literal(location)))

    # Create the company
    cnode = JB[company.lower().replace(' ', '_')]
    jg.add((vacancy, JOBS['vacancywith'], cnode))
    jg.add((cnode, RDF.type, JOBS['Organization']))

    # Ticker symbol
    if ticker != "":
        jg.add((cnode, COMPANY['symbol'], Literal(ticker)))
    jg.add((cnode, COMPANY['name'], Literal(company)))

    # Crunchbase (see also)
    if crunchbase != "":
        jg.add((cnode, RDFS['seeAlso'],
                Literal('http://api.crunchbase.com/v/1/company/%s.js' % \
                crunchbase)))

# Print the serialized graph
print jg.serialize(format='xml')
```

This code is very similar to what you've seen before—it simply reads the text file and asserts a set of triples in RDFLib for each line. It then creates an RDF file, which looks something like this:

```
<rdf:Description rdf:about="http://semprog.com/schemas/jobboard#1">
  <jobs:vacancywith rdf:resource="http://semprog.com/schemas/jobboard#metaweb"/>
  <rdf:type rdf:resource="http://www.medev.ac.uk/interoperability/rss/1.0/...
  <dc:title>Tugboat Captain</dc:title>
</rdf:Description>
...
  <rdf:Description rdf:about="http://semprog.com/schemas/jobboard#metaweb">
    <rdfs1:seeAlso>http://api.crunchbase.com/v/1/company/metawebtechnologies.js
    </rdfs1:seeAlso>
    <rdf:type rdf:resource="http://www.medev.ac.uk/interoperability/rss/1.0/...
    <company:name>Metaweb</company:name>
  </rdf:Description>
```

If for some reason you weren't able to create the RDF file, you can download it from *http://semprog.com/psw/chapter10/job_listings.rdf*. If this were a live application, you would likely receive updated data frequently, and you could set this up as a pipelined process that automatically converted all the tabular files to RDF as they arrived.

Loading the Data into Sesame

The easiest way to get the data into Sesame is through the user interface, uploading the data as you did in Chapter 8. If you have a Sesame instance running, feel free to upload the data file that way. You should be able to find a Sesame instance with all the data in this chapter loaded at *http://semprog.com:8180/openrdf-workbench* in the *chapter10* repository.

However, if you were constantly getting data from a source, you would want software to update it. Sesame lets you use PUT through its REST API to add RDF to a repository. To take advantage of this feature, we added an additional method to the connection object in pysesame for adding data. You can download the updated version from *http://semprog.com/psw/chapter10/pysesame.py*. Here's the new method:

```
def postdata(self,data):
    host = self.baseurl + '/repositories/' + self.repository + '/statements'
    res = urlopen(host, data)
    return res
```

This is pretty simple—it just uses HTTPLib to create a POST request and uploads the data in the body. It's very useful to have as part of the process. The great thing about this is that you don't even necessarily have to save the RDF to a file after it's generated; you can just put a couple of additional lines at the end of the code that generates it and have it add it right then:

```
data = jg.serialize(format='xml')
c = pysesame.connection('http://localhost:8280/openrdf-sesame/')
c.use_repository('joblistings')
c.postdata(data)
```

You can also use the POST method to add data to Sesame from within an application. A simple and common pattern is to build a small graph in memory from user input or an external source and add it to the repository. The other way to modify Sesame data is to use the RDFObject framework that we developed in Chapter 9.

Getting the data you have to a point where it can be queried is usually the first thing you'll want to do when building an application around semantic data. People coming from a tradition of dealing with relational databases are often tempted to design their whole ontology before even touching the data. Although ontologies are great for data sharing and for the RDFObject framework described in the previous chapter, you don't need to define the ontology before getting started.

Serving the Website

Since we're building a web application, we'll need an application server to query Sesame and a templating language to dynamically generate web pages. Since there are dozens of application servers to choose from, all with very extensive documentation, we'll just introduce a couple of components that we like and give a very quick overview of how they work. Presumably you already have your own favorite tools for building web applications, so we'll use only the most basic functionality to make it easy to port to a different application server.

CherryPy

CherryPy is a simple object-oriented application framework that is pretty low-level but very easy to use. It also provides a basic web server that you can use for testing, and which is perfect for the purposes of demonstrating how to build an application. (This way, you don't have to go through the hassle of configuring a real web server just to use these examples.)

You can find download and installation instructions on the CherryPy website (*http://cherrypy.org*), but if you have setuptools, you can just use `easy_install`:

```
$ easy_install cherrypy
```

Getting a web application running in CherryPy is very simple. All you need to do is define a Python class for the web application and methods within it representing the pages. Here's a basic example:

```
import cherrypy

class Main(object):

    @cherrypy.expose
    def index(self):
        return "Hello World!"

    @cherrypy.expose
    def greet(self,name):
        return "Hello, %s" % name

cherrypy.quickstart(Main())
```

The class `Main` is our application. It has two methods, `index` and `greet`, which represent two different pages in the application. Generally, the URL of a page is the method name, but `index` is a special case, as it's the root of the application. If you run this application and direct your browser to *http://localhost:8080/* (8080 is CherryPy's default port), you should see a page that says "Hello World!" CherryPy serves whatever is returned by the method.

If you point your browser to *http://localhost:8080/greet?name=Toby*, you'll see a page that says "Hello, Toby." We've passed in a parameter, which becomes a parameter to the method call. Both `index` and `greet` have the decorator `@cherrypy.expose` above them, which tells CherryPy that they can be requested as pages.

Mako Page Templates

Returning a small amount of text as the content of the page is simple enough within the CherryPy script, but real web applications usually return much more complex pages that would be very tedious to generate using just Python code. The usual solution is to use a templating language, which lets you create HTML pages with special code embedded in them. Unlike many web frameworks, CherryPy doesn't tell you which templating language to use, so you can install whichever one you like.

For this example we're going to use the Mako templating language, because it renders very quickly and is easy to understand. Like other templating languages, you use it by creating HTML files in a text or HTML editor and then adding special additional markup (in this case, embedded Python code and expressions). The Mako website can be found at *http://www.makotemplates.org/* and contains an introduction and download instructions. Again, if you have setuptools, you can install it with:

```
$ easy_install mako
```

We definitely encourage you to spend some time learning about Mako, but for now there are just a few things you need to know. Let's first look at the special constructs you can add to an HTML page to give it dynamic behavior. The most straightforward of these are the `<%` and `%>` tags, which simply let you embed regular Python code:

```
<%
    # names = ["Toby Segaran", "Jamie Taylor", "Colin Evans"]
    first_names = [name.split()[0] for name in names]
%>
```

This may look familiar to those of you who have used ASP or JSP before. The second construct is the use of `%` at the start of a line to embed control statements like `for` loops and `if-else` blocks. Here's a simple example:

```
<p>
    % if len(names) > 2:
        <strong>There are more than two names in the list</strong>
    % endif
</p>
```

Notice how we've introduced a non-Python expression, `endif`. This is necessary because Python usually indicates blocks with indentation, which doesn't really work with embedded code. Instead, Mako introduces the `endif` and `endfor` directives that specify when an `if` or `for` block ends, so the interpreter doesn't need to worry about the indentation.

The last template construction you'll need to know is `${<expression>}`, which lets you quickly embed an expression into the HTML. Here's an example:

```
This book was written by
<ul>
% for name in names:
    <li>${name}</li>
% endfor
</ul>
```

This will loop over all the names in the list and generate a list of them. You can put any valid Python expression inside the brackets—for example, we could have said `${name.upper()}` to show all the names in uppercase.

Using Mako within CherryPy just requires a couple extra lines of code. At the top of the file, we just import the necessary classes and tell Mako where it can find the templates:

```
from mako.template import Template
from mako.lookup import TemplateLookup

lookup = TemplateLookup(directories=['templates'])
```

We then initiate a `TemplateLookup` object with a list of directories in which to search for templates. From within an exposed method, we can use return statements to return rendered templates instead of just strings:

```
@cherrypy.expose
def index(self):
    authors = ['Toby Segaran','Jamie Taylor','Colin Evans']
    t = lookup.get_template('show_authors.html')
    return t.render(names=authors)
```

A template is retrieved using the `lookup` object and then rendered using a call to render. The render call takes as an argument a list of variables to be passed to the template. In this case, we're passing a list of three names in the variable `names`, which then allows us to use these names within the template, as in the examples just shown.

This quick primer on CherryPy and Mako should give you all the information you need to understand the web application in this chapter. So, let's get started!

A Generic Viewer

The first thing we want to build is a way to view the data in Sesame. The first viewer will be agnostic to the data, so we can have something that works right away, and then later we can build custom views for particular kinds of data. The viewer will show the properties of any object in the dataset and will allow us to click on those that are other objects to view them. Figure 10-2 shows what the final result will look like.

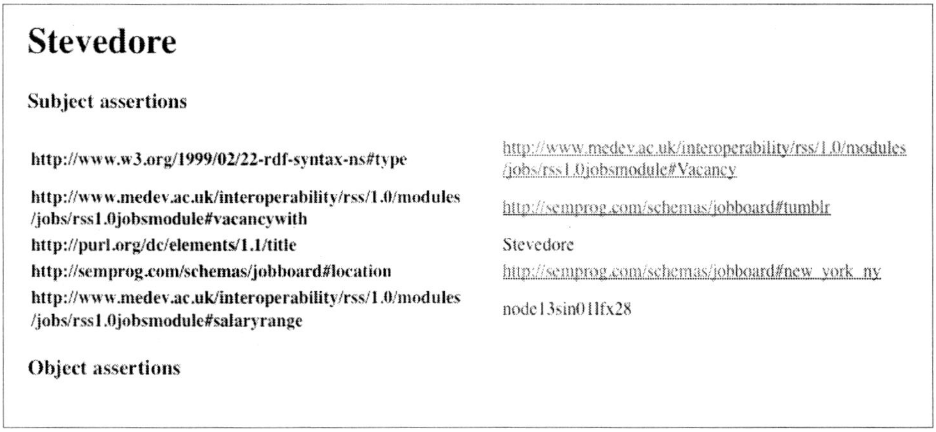

Figure 10-2. A generic viewer

To start with, let's set up a new CherryPy module that will be the basis of our application. The complete application file is available for you to download from *http://semprog.com/psw/chapter10/job_site.py*, or you can enter the code from this chapter as we go. Here's the basic setup:

```python
import cherrypy
from rdflib import Namespace, BNode, Literal, RDF, URIRef
from pysesame import connection
from mako.template import Template
from mako.lookup import TemplateLookup
from urllib import quote_plus

DC = Namespace("http://purl.org/dc/elements/1.1/")
JB = Namespace("http://semprog.com/schemas/jobboard#")
COMPANY = Namespace("http://purl.org/rss/1.0/modules/company/")
RDF = Namespace("http://www.w3.org/1999/02/22-rdf-syntax-ns#")

lookup = TemplateLookup(directories=['templates'])
con = connection('http://freerisk.org:8280/openrdf-sesame/')
con.use_repository('joblistings')
con.addnamespace('company', COMPANY)
con.addnamespace('rdf', RDF)

namefields=set([str(DC['title']), str(COMPANY['name'])])

class Main(object):

    @cherrypy.expose
    def index(self):
        id = quote_plus('http://semprog.com/schemas/jobboard#tumblr')
        return 'Try visiting <a href="view?id=%s">here</a>' % id

cherrypy.quickstart(Main())
```

Here we've just done the important imports, including the pysesame module from Chapter 8. We've also created a few empty exposed methods that we'll be filling in throughout the chapter.

Getting Data from Sesame

The first step in building a viewer is to get the data from Sesame, which we'll do in the exposed `view` method using the pysesame connection that we've already established. The `view` method has a single parameter, `id`, in which we pass the ID of the object whose triples we want to view. We want to retrieve all the triples that have this ID as the subject or the object (i.e., where this ID appears in the first or third spot). Take a look at this simplified `view` method:

```
@cherrypy.expose
def view(self,id):
    sa = con.query('select ?pred ?obj where {<%s> ?pred  ?obj .}' % id)
    oa = con.query('select ?pred ?sub where {?sub ?pred  <%s> .}' % id)
    name = id
    for row in sa:
        print row
        if row['pred']['value'] in namefields:
            name = row['obj']['value']

    t = lookup.get_template('viewgeneric.html')
    return t.render(name=name, id=id, sa=sa, oa=oa, qp=quote_plus)
```

Using the connection, the method performs two different queries: one to find triples where this ID is the object, and the other to find triples where the ID is the subject. The method divides these results into things that refer to literal values and things that refer to other objects. It also pulls out the title or name if there's one that it recognizes. It then grabs the Mako template for the generic view, passes it the results of the query, and returns the rendered template, as you saw earlier in Figure 10-2.

The Generic Template

The template for the basic view just needs to display the literal values of the object we're looking at in a clean, attractive manner, and then links to all the objects to which it's connected. Our `view` method passes three objects to the template: `literal_triples`, `subject_triples`, and `object_triples`. To display them, we create a template that looks like this (you can download it from *http://semprog.com/psw/chapter10/viewgeneric .html*):

```
<html>
    <head>
        <title>Viewing ${name}</title>
    </head>
    <body>
        <h1>${name}</h1>
        <h3>Subject assertions</h3>
```

```
<table>
% for row in sa:
    <tr>
        <td style="font-weight:bold">${row['pred']['value']}</td>
        % if row['obj']['type'] == 'uri':
        <td><a href="view?id=${qp(row['obj']['value'])}">
                ${row['obj']['value']}
            </a></td>
        % else:
        <td>${row['obj']['value']}</td>
        % endif
    </tr>
% endfor
</table>
<h3>Object assertions</h3>
<table>
% for row in oa:
    <tr>
        <td style="font-weight:bold">${row['pred']['value']}</td>
        % if row['sub']['type'] == 'uri':
        <td><a href="view?id=${qp(row['sub']['value'])}">
            ${row['sub']['value']}
        </a></td>
        % else:
        <td>${row['sub']['value']}</td>
        % endif
    </tr>
% endfor
</table>
</body>
</html>
```

Notice how the header on the page is displayed as the name, if it has one; otherwise, the header is just the ID. If you're building the application as you go through the chapter, put this file in the templates subdirectory of your CherryPy application. Run the application and point your browser to *http://localhost:8080/view?id=job:8301283* to see it in action. You should be able to click around to other objects, such as places and companies, but there won't be any information about them except other job listings.

Getting Company Data

We now actually have a working application, but it's not much fun yet. Pages that aren't about jobs don't really tell us much because we don't have any other data. However, we're in a good position—if we can find a data source, then not only we can start adding any triples we like to Sesame, but they'll also automatically display in our generic viewer. In this section, we'll look at some sources of data and put a few more company facts into our data store.

Crunchbase

When looking for additional sources of data, it is important to seek out those that can be easily parsed, but even more importantly those that use strong keys for as many different records as possible. One of the best databases for information about startup companies that fits these requirements is Crunchbase, created by the company Techcrunch and maintained by a community of users. It has strong keys, referred to as *permalinks* in their nomenclature, and is completely accessible through a JavaScript API.

Crunchbase has a very nice REST API, which takes simple GET requests and returns JSON objects. It allows you to access data about companies, people, products, and financial organizations. For example, if you point your browser to *http://api.crunchbase .com/v/1/company/metawebtechnologies.js*, you'll see a JSON object describing everything Crunchbase knows about Metaweb:

```
{"name": "Metaweb Technologies",
 "permalink": "metawebtechnologies",
 "crunchbase_url": "http://www.crunchbase.com/company/metawebtechnologies",
 "homepage_url": "http://www.metaweb.com/",
 ...
 "products":
 [{"name": "freebase",
   "permalink": "freebase"}],
 ...
 "funding_rounds":
 [{"round_code": "b",
   "source_url": "http://www.paidcontent.org/entry/419-metaweb-gets-42-...
   "source_description": "MetaWeb Gets $42 Million Second Round For Online Database",
   "raised_amount": 42000000.0,
   "raised_currency_code": "USD",
   "funded_year": 2008,

   ...
```

This data provides useful additional information about companies in our current dataset. Because some of the data we have in Sesame already has Crunchbase keys, we can easily construct the URLs to get the additional data, use a JSON parser to extract fields that are interesting, and create new triples about the companies. Here's a simple script called *crunchbase_data.py* (download from *http://semprog.com/psw/chapter10/ crunchbase_data.py*) that adds some key pieces of information to the companies in Sesame:

```
from rdflib.Graph import ConjunctiveGraph
from rdflib import Namespace, BNode, Literal, RDF, URIRef
from pysesame import connection
import urllib
from simplejson import loads

# Connection to Sesame
con = connection('http://freerisk.org:8280/openrdf-sesame/')
con.use_repository('joblistings')
```

```
con.addnamespace('rdfs', 'http://www.w3.org/2000/01/rdf-schema#')

JB = Namespace("http://semprog.com/schemas/jobboard#")
DC = Namespace("http://purl.org/dc/elements/1.1/")
FOAF = Namespace("http://xmlns.com/foaf/0.1/")

cg = ConjunctiveGraph()
cg.bind('dc', DC)
cg.bind('jobboard', JB)
cg.bind('foaf', FOAF)

# Find seeAlso URLs containing Crunchbase
res=con.query('select ?id ?url where '+\
              '{?id rdfs:seeAlso ?url . FILTER regex(?url, "crunchbase")}')

# Loop over the results
for row in res:
    company = URIRef(row['id']['value'])
    url = row['url']['value']
    data = urllib.urlopen(url).read()
    record = loads(data)

    # Add company locations
    for loc in record['offices']:
        if loc['city'] and loc['state_code']:
            cityid = '%s_%s' % (loc['city'].lower().\
                    replace(' ','_'), loc['state_code'].lower())
            cg.add((company, JB['location'], JB[cityid]))

    # Add URL
    cg.add((company, FOAF['homepage'], Literal(record['homepage_url'])))

    # Add funding rounds
    for fr in record['funding_rounds']:
        round = BNode()
        cg.add((company, JB['funding_round'], round))
        cg.add((round, JB['amount'], Literal(fr['raised_amount'])))
        cg.add((round, DC['date'], Literal(str(fr['funded_year']))))

# Upload graph to Sesame
data= jg.serialize(format='xml')
c = pysesame.connection('http://freerisk.org:8280/openrdf-sesame/')
c.use_repository('joblistings')
print c.postdata(data)
```

This script begins with a query that finds all the companies in the set that have Crunchbase keys, and for each of these companies it requests and parses the relevant Crunchbase API page. To keep our script short, we're only interested in a few fields here—the URLs, company locations, and fund-raising amounts—so for each of these it checks to see if there's already a matching triple, and if not, it creates it. Notice that the funding round is created as a separate object so that it can have a lot of additional data. Figure 10-3 shows what this looks like as a graph.

Company
funding_round
date
date
homepage
location
Funding round
amount
uri
float
City

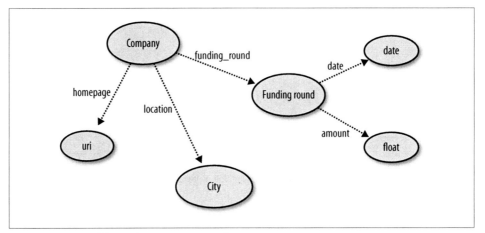

Figure 10-3. Crunchbase adaption to our graph

If you run the script, it should add fund-raising and location information to the companies. Going back to your browser, the company pages should now have more information, as shown in Figure 10-4.

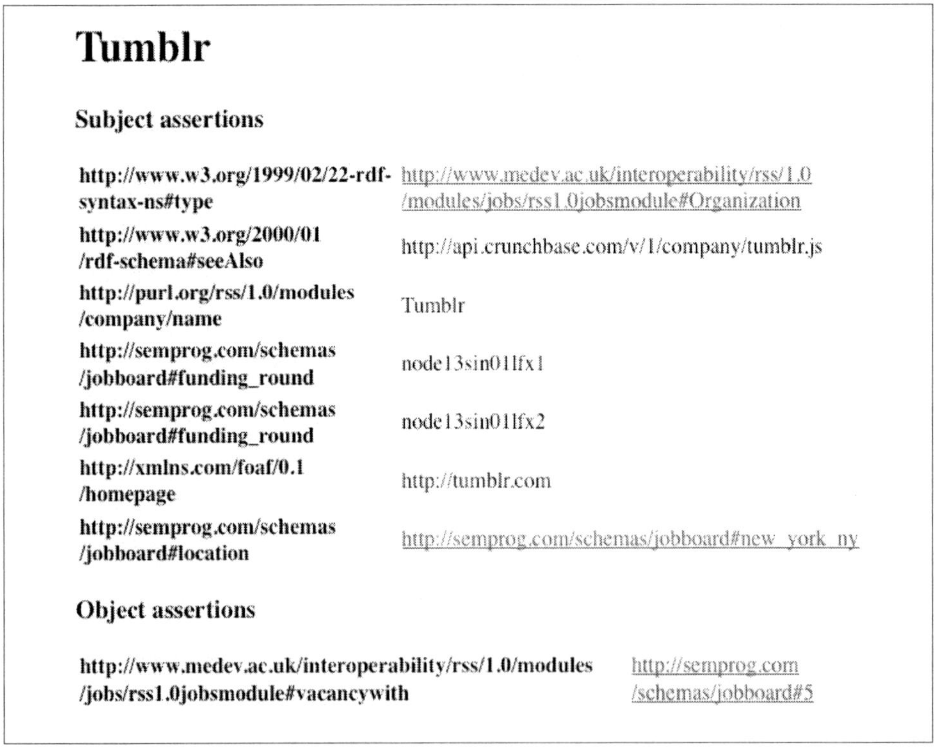

Tumblr
Subject assertions
http://www.w3.org/1999/02/22-rdf-syntax-ns#type
http://www.medev.ac.uk/interoperability/rss/1.0/modules/jobs/rss1.0jobsmodule#Organization
http://www.w3.org/2000/01/rdf-schema#seeAlso
http://api.crunchbase.com/v/1/company/tumblr.js
http://purl.org/rss/1.0/modules/company/name
Tumblr
http://semprog.com/schemas/jobboard#funding_round
node13sin01lfx1
http://semprog.com/schemas/jobboard#funding_round
node13sin01lfx2
http://xmlns.com/foaf/0.1/homepage
http://tumblr.com
http://semprog.com/schemas/jobboard#location
http://semprog.com/schemas/jobboard#new_york_ny
Object assertions
http://www.medev.ac.uk/interoperability/rss/1.0/modules/jobs/rss1.0jobsmodule#vacancywith
http://semprog.com/schemas/jobboard#5

Figure 10-4. Company generic view with more data added

Yahoo! Finance

Crunchbase is a great source of data about young tech companies, but for large, publicly traded companies there are many better sources. These larger companies are usually required to submit a lot of data to the government of the country in which they're traded, and a lot of sites aggregate this information and make it easily accessible. One such site is Yahoo! Finance, shown in Figure 10-5.

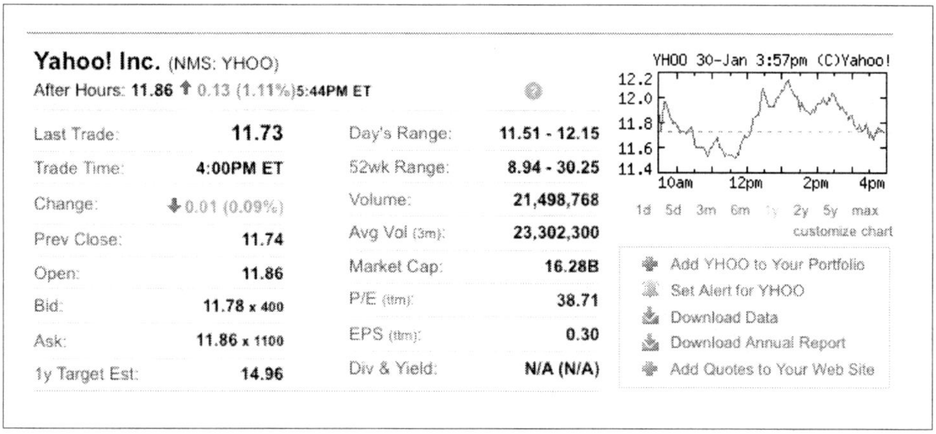

Figure 10-5. Yahoo! Finance

Yahoo! Finance doesn't offer an API, and we're not going to get into web-scraping here—for that, you may want to check out Toby's book *Programming Collective Intelligence (http://oreilly.com/catalog/9780596529321/)*—but it does offer an easy way to download recent stock quotes as a text file. A GET request to a URL such as:

```
http://finance.yahoo.com/q/hp?s=YHOO&a=00&b=28&c=2008&d=00&e=28&f=2009&g=m
```

returns a list of stock prices, in this case monthly quotes for Yahoo! from January 2008 to January 2009:

```
Date,Open,High,Low,Close,Volume,Adj Close
2009-01-02,12.17,13.56,10.81,11.34,21178500,11.34
2008-12-01,11.82,13.57,10.50,12.20,20089400,12.20
...
2008-03-03,27.73,29.18,25.72,28.93,30491900,28.93
2008-02-01,28.68,30.25,27.34,27.78,67246000,27.78
2008-01-28,21.56,21.90,18.58,19.18,77649000,19.18
```

Since we're building a job listing application, actual stock prices aren't necessarily very interesting to us. However, the recent performance of a company's stock price might be some indication of how well that company is doing (perhaps it's a stretch, but maybe you'll be working with people who are happy about their stock options!). So what we'll do is add a 12-month change in stock price to companies in our database that have a ticker symbol. You can find the following script at *http://semprog.com/psw/chapter10/yahoo_finance_update.py*:

```python
from rdflib.Graph import ConjunctiveGraph
from rdflib import Namespace, BNode, Literal, RDF, URIRef
from pysesame import connection
import urllib
from simplejson import loads

# Connection to Sesame
con = connection('http://freerisk.org:8280/openrdf-sesame/')
con.use_repository('joblistings')
con.addnamespace('rdfs','http://www.w3.org/2000/01/rdf-schema#')

JB = Namespace("http://semprog.com/schemas/jobboard#")
DC = Namespace("http://purl.org/dc/elements/1.1/")
FOAF = Namespace("http://xmlns.com/foaf/0.1/")

cg=ConjunctiveGraph()
cg.bind('dc', DC)
cg.bind('jobboard', JB)
cg.bind('foaf', FOAF)

# Find seeAlso URLs containing Crunchbase
res = con.query('select ?id ?url where {?id rdfs:seeAlso ?url . '+\
                'FILTER regex(?url, "crunchbase")}')

# Loop over the results
for row in res:
    company = URIRef(row['id']['value'])
    url = row['url']['value']
    data = urllib.urlopen(url).read()
    record = loads(data)

    # Add company locations
    for loc in record['offices']:
        if loc['city'] and loc['state_code']:
            cityid = '%s_%s' % (loc['city'].lower().replace(' ', '_'), \
                     loc['state_code'].lower())
            cg.add((company,JB['location'],JB[cityid]))

    # Add URL
    cg.add((company, FOAF['homepage'], Literal(record['homepage_url'])))

    # Add funding rounds
    for fr in record['funding_rounds']:
        round = BNode()
        cg.add((company, JB['funding_round'], round))
        cg.add((round, JB['amount'], Literal(fr['raised_amount'])))
        cg.add((round, DC['date'], Literal(str(fr['funded_year']))))


data = jg.serialize(format='xml')
c = pysesame.connection('http://freerisk.org:8280/openrdf-sesame/')
c.use_repository('joblistings')
print c.postdata(data)
```

This makes another nice addition to our database, which you'll now see in the generic viewer when looking at a public company. If you like, you can try to get more data about public companies from Yahoo! Finance, some other finance site, or even the Securities and Exchange Commission website.

Reconciling Freebase Connections

Our initial table didn't have ticker symbols or Crunchbase keys for every company, which meant that we couldn't get additional data from one or both of our sources. *Reconciliation*, the process of matching keys in one database to keys in another database, is an ongoing area of research in both traditional and semantic data integration.

Freebase, which we introduced in Chapter 5, has mapped a lot of keys from various databases onto common objects, including Crunchbase IDs, ticker symbols, Wikipedia keys, and SEC numbers. We can use these mappings to add more keys to our Sesame store and thus get even more data from various sources.

The following script, which you can find at *http://semprog.com/psw/chapter10/freebase_companies.py*, searches Freebase for companies by name and by all the keys it has to see if it can find additional keys:

```
from rdflib.Graph import ConjunctiveGraph
from rdflib import Namespace, BNode, Literal, RDF, URIRef
from pysesame import connection
import metaweb

COMPANY = Namespace("http://purl.org/rss/1.0/modules/company/")
JOBS = Namespace("http://www.medev.ac.uk/interoperability/rss/1.0/"+\
                "modules/jobs/rss1.0jobsmodule#")
RDFS = Namespace('http://www.w3.org/2000/01/rdf-schema#')

# Connection to Sesame
con = connection('http://freerisk.org:8280/openrdf-sesame/')
con.use_repository('joblistings')
con.addnamespace('jobs', str(JOBS))
con.addnamespace('company', str(COMPANY))
con.addnamespace('rdf', 'http://www.w3.org/1999/02/22-rdf-syntax-ns#')

cg = ConjunctiveGraph()
cg.bind('company', COMPANY)

# Find companies with ticker symbols
res = con.query('select ?id ?ticker where {?id rdf:type jobs:Organization .'+
                                '?id company:symbol ?ticker .}')

for row in res:
    company = URIRef(row['id']['value'])
    ticker = row['ticker']['value']
    mq = metaweb.read({'type':'/business/company',
                      'key':{'namespace':'/authority/crunchbase/company',
                      'value':None},
                      'ticker_symbol':{'ticker_symbol':ticker}})
```

```
    if mq != None:
        cg.add((company,RDFS['seeAlso'],
                Literal('http://api.crunchbase.com/v/1/company/%s.js' % \
                        mq['key']['value'])))

# Now you do the reverse!
data = jg.serialize(format='xml')
c = pysesame.connection('http://freerisk.org:8280/openrdf-sesame/')
c.use_repository('joblistings')
print c.postdata(data)
```

If necessary, you can refer back to Chapter 5 to understand the MQL query used here. For each company that has a ticker symbol, the query uses Freebase to try to find the matching Crunchbase key. It then adds these keys as **seeAlso** links to Sesame. As an exercise, see if you can make the query do the reverse (find companies with Crunchbase keys and add their ticker symbols). After you run this script, you can run the Yahoo! Finance and Crunchbase scripts again to get even more data into your database.

This particular set of scripts is a nice example of the rule-chaining pattern that we demonstrated in Chapter 3. Our Freebase script adds more keys to the database, which allows our other scripts to find more information. In theory, we can keep piling on more and more scripts that add keys and data to the database from various sources and that help each other out without even being aware of each other. This type of loose coupling around data can be very clean and powerful.

Specialized Views

Our generic viewer was a great way to get started and explore new types of data, but now that the direction of the application is beginning to become clearer, we can start building special views for the different types of objects. Semantic data makes it very easy to think about everything as just assertions, but in reality the way people want to see information about a job posting can be quite different from the way they want to look at a company, a person, or a place.

The first thing we'll do here is modify the `view` method so that it looks to see if there's a special viewer for the type of object we're trying to view:

```
@cherrypy.expose
def view(self,id):

    # Get the list of types for this object
    tq = con.query('select ?type where {<%s> rdf:type ?type .}' % id)
    ob_types = [t['type']['value'] for t in tq]

    # Look to see if there's a special handler
    for typename, func in self.type_views:
        if str(typename) in ob_types:
            return func(self, id)
```

```
    # If not, use the generic view
    sa = con.query('select ?pred ?obj where {<%s> ?pred  ?obj .}' % id)
    oa = con.query('select ?pred ?sub where {?sub ?pred  <%s> .}' % id)
    name = id
    for row in sa:
        print row
        if row['pred']['value'] in namefields:
            name = row['obj']['value']

    t = lookup.get_template('viewgeneric.html')
    return t.render(name=name, sa=sa, oa=oa, qp=quote_plus)

type_views = [(JOBS['Organization'],view_company),
              (JOBS['Vacancy'],view_job)]
```

The view method looks through the list of types to see if one matches and, if so, calls the indicated view method. If no matches are found, it defaults to the generic viewer. Unlike the usual objects in object-oriented systems, semantic objects can have multiple types. Because of this, the view method has an order of preference of views.

Let's add a couple of alternative view methods to the list. Notice that we can extend the queries further and actually use generated objects instead of just passing triples to the template:

```
def view_job(self,id):
    # Get all job details
    job = con.query("""select ?title ?salary ?cid ?cname
                    where {<%s> dc:title ?title .
                           <%s> jobs:vacancywith ?cid .
                           ?cid company:name ?cname .
                           <%s> jobs:salaryrange ?sr .
                           ?sr jobs:minimumsalary ?salary .}""" % (id, id, id))
    if len(job) == 0: return "No job found with that ID"

    t = lookup.get_template('viewjob.html')
    return t.render(title=job[0]['title']['value'],
                    salary=job[0]['salary']['value'], cid=job[0]['cid']['value'],
                    cname=job[0]['cname']['value'], qp=quote_plus)

def view_company(self, id):
    # get the name
    nq = con.query('select ?name where {<%s> company:name ?name}' % id)
    if len(nq)>0: name = nq[0]['name']['value']
    else: name = 'Unknown'

    # get the homepage
    homeq = con.query('select ?url where {<%s> foaf:homepage ?url .}' % id)
    if len(homeq) > 0: homepage=homeq[0]['url']['value']
    else: homepage = None

    # get locations
    locations = con.query('select ?loc where {<%s> jb:location ?loc .}' % id)
```

```
        # get job listings
        jobs = con.query("""select ?title ?job
                        where {?job jobs:vacancywith <%s> .
                                ?job dc:title ?title}""" % id)

        # get funding rounds
        funding = con.query("""select ?amount ?date
                          where {<%s> jb:funding_round ?round .
                                ?round jb:amount ?amount .
                                ?round dc:date ?date .}""" % id)
        t = lookup.get_template('viewcompany.html')
        return t.render(name=name, jobs=jobs, funding=funding, locations=locations,
                    homepage=homepage, qp=quote_plus)
```

We've referred to a couple of new templates in these **view** methods. First let's create a new template for job postings (we encourage you to create a prettier one than this—space limitations prevent us from going too crazy on the CSS!), called *job.html*:

```
<html>
    <head>
        <title>Job Listing</title>
    </head>
    <body>
        <h2>Job listing</h2>
        <table cellpadding="4">
            <tr>
                <td style="font-weight:bold;">Title</td>
                <td>${title}</td>
            </tr>
            <tr>
                <td style="font-weight:bold;">Company</td>
                <td><a href="view?id=${qp(cid)}">${cname}</a></td>
            </tr>
            <tr>
                <td style="font-weight:bold;">Salary</td>
                <td>$${salary}</td>
            </tr>
        </table>
    </body>
</html>
```

And now another template for companies, *company.html*:

```
<html>
    <head>
        <title>${name}</title>
        <style>
            #main {width:680px;}
            #joblist {float:left;}
            #company_info {float:right;width:240px;
                        background-color:#E0E0E0;padding-left:10px}
            h3 {margin-bottom: 5px;}
        </style>
    </head>
    <body>
```

```
<h1>Company Information: ${name}</h1>
<div id="main">
    <div id="company_info">
        <h2>Company details</h2>
        % if homepage != None:
            <a href="${homepage}">${homepage}</a>
        % endif
        <h3>Funding</h3>
        <table cellpadding="3">
            <tr><th>Date</th><th>Amount</th></tr>
            % for f in funding:
                <tr><td>
                  ${f['date']['value']}</td><td>$$${f['amount']['value']}
                </td></tr>
            % endfor
        </table>
        <h3>Locations</h3>
        <ul>
        % for l in locations:
            <li>
                <a href="view?id=${qp(l['loc']['value'])}">
                    ${l['loc']['value'].split('#')[1]}
                </a>
            </li>
        % endfor
        </ul>
    </div>
    <div id="joblist">
        <h2>Current job listings</h2>
        <ul>
        % for job in jobs:
            <li><a href="view?id=${qp(job['job']['value'])}">
                ${job['title']['value']}
            </a></li>
        % endfor
        </ul>
    </div>
</div>
    </body>
</html>
```

You can download both of these from *http://semprog.com/psw/chapter10/templates*. Now if you view a company, instead of the generic view you should see something like Figure 10-6.

Isn't that much better? Also notice that if you click on a location, you still get the old generic view. This is great for development, as it lets you expand the kinds of data you have without immediately having views for it. This can make testing and prototyping much simpler.

Figure 10-6. Screenshot of special "company" view

Publishing for Others

A key philosophical lesson we hope you take from this book is that there is really no distinction between data providers and data consumers. When you build an application that puts a lot of information online, some of which is from other sources and some of which is original, it's better if that information is accessible to others not just as text but in a machine-readable form as well. When it's easy to extract the important fields from the information, it allows other people to remix it in their own way. Although this all may sound a little utopian, it's clear that many companies have benefited greatly from their use of open APIs.

In our case, we're not interested in inventing a new API just for our data. We can use semantic web standards to republish the data in a way that's easy for others to consume.

RDFa

In Chapter 7 we introduced RDFa, which is a way of embedding semantics in regular web pages. Because all our data is generated from semantics anyway, it's easy enough to alter the templates to include RDFa that crawlers can use to understand the content of the page.

As a simple example, let's update the generic view template to include RDFa, generated from the triples that were passed to it:

```
<html xmlns="http://www.w3.org/1999/xhtml">
    <head>
        <title>Viewing ${name}</title>
    </head>
    <body>
        <h1>${name}</h1>
        <h3>Subject assertions</h3>
        <table about="${rid}">
        % for row in sa:
            <tr>
                <td style="font-weight:bold">${row['pred']['value']}</td>
                % if row['obj']['type'] == 'uri':
                <td rel="${row['pred']['value']}" resource="${row['obj']['value']}">
                    <a href="view?id=${qp(row['obj']['value'])}">
                        ${row['obj']['value']}
                    </a>
                </td>
                % else:
                <td property="${row['pred']['value']}">${row['obj']['value']}</td>
                % endif
            </tr>
        % endfor
        </table>
        <h3>Object assertions</h3>
        <table>
        % for row in oa:
            <tr>
                <td style="font-weight:bold">${row['pred']['value']}</td>
                % if row['sub']['type'] == 'uri':
                <td about="${row['sub']['value']}">
                    <a href="view?id=${qp(row['sub']['value'])}">
                        ${row['sub']['value']}
                    </a>
                    <span rel="${row['pred']['value']}" resource="${rid}"/>
                </td>
                % else:
                <td>${row['sub']['value']}</td>
                % endif
            </tr>
        % endfor
        </table>
    </body>
</html>
```

Now if you go to a page that still uses the generic view (such as the page for New York) and view the HTML source, you'll see the RDFa embedded in the tags:

```
<tr>
    <td style="font-weight:bold">http://semprog.com/schemas/jobboard#location</td>
    <td about="http://semprog.com/schemas/jobboard#google">
        <a href="view?id=http%3A%2F%2Fsemprog.com%2Fschemas%2Fjobboard%23google">
            http://semprog.com/schemas/jobboard#google
        </a>
        <span rel="http://semprog.com/schemas/jobboard#location"
              resource="http://semprog.com/schemas/jobboard#new_york_ny"/>
```

```
    </td>
  </tr>
```

See if you can do the same for the specialized views. The biggest downside to embedding data within the page is that each time you change the way the data is viewed, you have to make sure that the embedded content is still present and correct. This makes sense, because both RDFa and microformats were designed to be easy for people to include in arbitrary web pages. When the page is generated automatically, it is often easier to separate the machine-readable data from the user-readable data.

RDF/XML

An alternative way to publish the data in a machine-readable form is to have our `view` method return RDF/XML instead of rendered HTML in the right circumstances. The first step is defining exactly those circumstances and being able to detect them. Consider the following code snippet:

```
def view(self,id,format=None):
    if format == None:
        if 'Accept' in cherrypy.request.headers:
            accept = cherrypy.request.headers['Accept'].split(',')
            if 'application/rdf+xml' in accept:
                format='rdf'
```

There are two different ways for the server to determine that the request is for RDF/ XML data. The simplest is that the requester makes it explicit by adding `format=rdf` to the request, which might be similar to other REST APIs you've seen. You can make it clear to crawlers that this RDF version of the page is available by adding a line like this to the templates:

```
<link rel="alternate" type="application/rdf+xml"
 href="http://localhost:8080/view?id=xxx&format=rdf" />
```

However, if no format is specified in the request, the method attempts to determine if the requester wants RDF/XML back as a result by looking at the HTTP request headers provided by CherryPy. This is preferable, as it means that crawlers looking for RDF just get back the machine-readable form of the data right away. Most web frameworks allow you to look at the request headers and return data in the right format. It's actually a good idea to have both methods of retrieving the RDF available, so that if people want to look at RDF in their browser for debugging purposes or just out of curiosity, they can do so simply by changing the URL.

To return RDF instead of a rendered HTML page, we can just use a SPARQL CONSTRUCT query to pull out all the triples and return them directly, no template required! Here's the beginning of the new `view` method:

```
@cherrypy.expose
def view(self, id, format=None):
    if format == None:
        if 'Accept' in cherrypy.request.headers:
```

```
            accept = cherrypy.request.headers['Accept'].split(',')
            if 'application/rdf+xml' in accept:
                format = 'rdf'

        if format == 'rdf':
            data = con.construct_query("""construct {?a ?b ?c .}
                                          where {{?a ?b ?c} .
                                          {{<%s> ?b ?c} UNION
                                           {?a ?b <%s>}} .}""" % (id, id))
            return data

        # Get the list of types for this object
        ... etc ...
```

In a perfect world, you'd be able to publish everything with microformats, RDFa, and RDF/XML so that any other program would be able to consume your data no matter what it was looking for. However, if you're just looking for an easy way to publish to a standard, you can almost always just apply the few ideas in this section and have RDF/XML online quickly.

Expanding the Data

If there's one message we hope we've beaten into you by now, it's that semantic data is *flexible*—it always leaves you the option to add more, making your applications more compelling and your queries more sophisticated. In this section, we're going to prove it by adding more data to the cities where the jobs and companies are located (which, as of right now, are pretty empty).

We'll add data about the geographic locations of the cities and then about the demographics of the cities. We'll do this without messing with schema migration, and later we'll use the data we added to make some interesting queries.

Locations

The geolocations—the longitudes and latitudes—are the first thing we'll add to the cities. You can find geolocations in a lot of different places; there are tables on the Web, and they can also can be found in Freebase. For this example, we've provided a list of cities along with their geolocations at *http://semprog.com/psw/chapter10/city_locations .csv*. It's a simple comma-delimited file that looks like this:

```
san_francisco_ca,37.775,-122.4183
new_york_ny,40.7142,-74.0064
boulder_co,40.015,-105.27
pittsburgh_pa,40.4406,-79.9961
...
```

By now, you should know the drill. Parse the data, convert it to an RDF graph, and load it into Sesame. Here's a script, which you can find at *http://semprog.com/psw/ chapter10/load_city_locations.py*:

```
from rdflib.Graph import ConjunctiveGraph
from rdflib import Namespace, BNode, Literal, RDF, URIRef
import csv
import pysesame

JB = Namespace("http://semprog.com/schemas/jobboard#")
GEO = Namespace('http://www.w3.org/2003/01/geo/wgs84_pos#')

lg = ConjunctiveGraph()
lg.bind('geo', GEO)

for city, lat, long in csv.reader(file('city_locations.csv', 'U')):
    lg.add((JB[city], GEO['lat'], Literal(float(lat))))
    lg.add((JB[city], GEO['long'], Literal(float(long))))

data = lg.serialize(format='xml')
print data
c = pysesame.connection('http://freerisk.org:8280/openrdf-sesame/')
c.use_repository('joblistings')
print c.postdata(data)
```

Note that we've used the geolocation schema from the W3C, which you can find at *http://www.w3.org/2003/01/geo/*. It defines the "lat" and "long" predicates that we apply to the cities in Sesame.

Geography, Economy, Demography

We're also interested in things like the population, area, and average cost of rent in a city. Because you're surely tired of reading parser code by now, we've provided an RDF file of demographic data at *http://semprog.com/psw/chapter10/demographics.rdf*. Here's a short excerpt (including a new namespace for census data):

```
<?xml version="1.0" encoding="UTF-8"?>
<rdf:RDF
   xmlns:census="tag:govshare.info,2005:rdf/census/"
   xmlns:dc="http://purl.org/dc/elements/1.1/"
   xmlns:jobboard="http://semprog.com/schemas/jobboard#"
   xmlns:rdf="http://www.w3.org/1999/02/22-rdf-syntax-ns#"
>
  <rdf:Description rdf:about="http://semprog.com/schemas/jobboard#tempe_az">
    <census:population rdf:datatype="http://www.w3.org/2001/XMLSchema#int">
       174091
    </census:population>
    <jobboard:onebedrent rdf:datatype="http://www.w3.org/2001/XMLSchema#int">
       727
    </jobboard:onebedrent>
    <dc:title>Tempe</dc:title>
    <census:landarea rdf:datatype="http://www.w3.org/2001/XMLSchema#float">
       40.2
    </census:landarea>
  </rdf:Description>
...
```

The easiest way to load this data is to simply enter the URL in the "Add data" page of the Sesame workbench. If you'd rather download it and then upload it to Sesame, that will work too. Figure 10-7 shows the new generic view of New York.

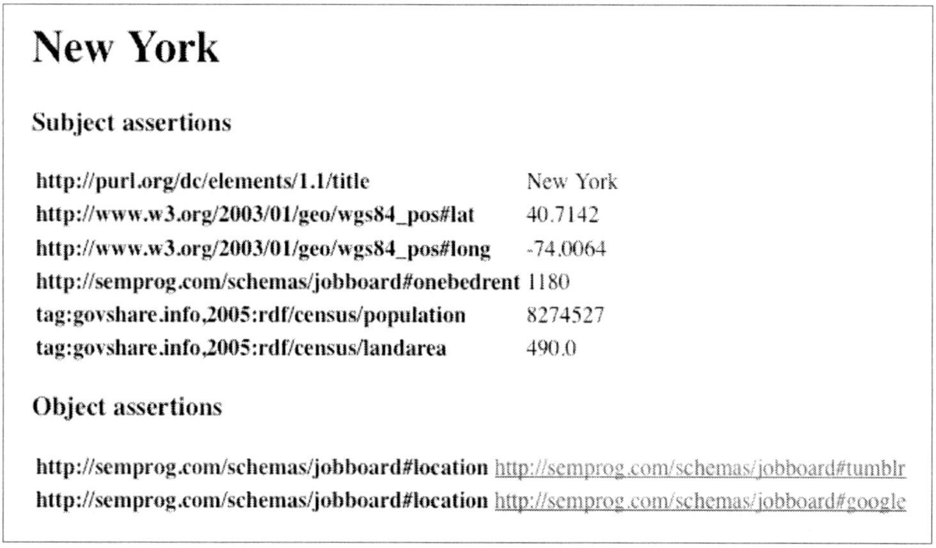

New York

Subject assertions

http://purl.org/dc/elements/1.1/title	New York
http://www.w3.org/2003/01/geo/wgs84_pos#lat	40.7142
http://www.w3.org/2003/01/geo/wgs84_pos#long	-74.0064
http://semprog.com/schemas/jobboard#onebedrent	1180
tag:govshare.info,2005:rdf/census/population	8274527
tag:govshare.info,2005:rdf/census/landarea	490.0

Object assertions

http://semprog.com/schemas/jobboard#location	http://semprog.com/schemas/jobboard#tumblr
http://semprog.com/schemas/jobboard#location	http://semprog.com/schemas/jobboard#google

---

Figure 10-7. Screenshot of New York with lots more data added

Sophisticated Queries

One of the best reasons for linking many different kinds of data together is to be able to ask complex questions that you wouldn't have been able to ask before. We've linked together data about company investments, stock performance, job listings, and city demographics, so we should now have the ability to ask questions that cross all these domains.

First of all, let's make a template for displaying the results of a job query. It'll take the JSON results of a SPARQL query and display the job title, salary, and location. Here's *jobquery.html* (available at *http://semprog.com/psw/chapter10/templates/jobquery .html*):

```
<html>
    <head>
        <title>Job query results</title>
    </head>
    <body>
        <h2>Job query results</h2>
        <table cellpadding="5">
        <tr>
            <th>Job</th><th>Company</th><th>Salary</th><th>Location</th>
        </tr>
        % for job in jobs:
```

```
<tr>
    <td>
        <a href="view?id=${qp(job['job']['value'])}">
            ${job['title']['value']}
        </a>
    </td>
    <td>
        <a href="view?id=${qp(job['cid']['value'])}">
            ${job['cname']['value']}
        </a>
    </td>
    <td>$${job['salary']['value']}</a></td>
    <td>${job['location']['value']}</a></td>
</tr>
% endfor
</table>
</body>
</html>
```

We can reuse this template and write some methods that do fun queries. For a first example, let's say you're only interested in applying for a job at companies whose stock price hasn't fallen much recently. You could create a method like this one:

```
@cherrypy.expose
def jobs_price_change(self, min=0.0):
    jobs = con.query("""select ?job ?title ?salary ?location ?cid ?cname
                        where {?job jobs:vacancywith ?cid .
                               ?cid company:name ?cname .
                               ?cid jb:stockpricechange ?change .
                               filter(?change>=%f)
                               ?job dc:title ?title .
                               ?job jobs:salaryrange ?range .
                               ?range jobs:minimumsalary ?salary .
                               ?job jb:location ?location . }""" % float(min))
    t = lookup.get_template('jobquery.html')
    return t.render(jobs=jobs,qp=quote_plus)
```

This takes a parameter for the stock price cutoff and uses it in a SPARQL query that queries across jobs and companies. Since it's an exposed method, you can go straight to it in your browser. Try *http://localhost:8080/jobs_price_change?min=-60* (it's been a bad market!), and you should get something like Figure 10-8.

Job query results

Job	Company	Salary	Location
Third Mate	Yahoo	$40000.0	http://semprog.com/schemas/jobboard#tempe_az

Figure 10-8. Job query results from a stock price query

How about a job containing a certain keyword in a city with at least a given population?

```
@cherrypy.expose
def job_keyword_pop(self,keyword='',pop=0):
    jobs = con.query("""select ?job ?title ?salary ?location ?cid ?cname
                        where {?job jobs:vacancywith ?cid .
                            ?cid company:name ?cname .
                            ?job dc:title ?title . filter regex(?title,"%s")
                            ?job jobs:salaryrange ?range .
                            ?range jobs:minimumsalary ?salary .
                            ?job jb:location ?location .
                            ?location census:population ?pop .
                            filter (?pop>%d)}""" % (keyword, int(pop)))

    print jobs
    t = lookup.get_template('jobquery.html')
    return t.render(jobs=jobs, qp=quote_plus)
```

And if you're really picky, maybe you want to find a job in a small city that pays at least some multiple of the average rent it costs to live in that city, at a company that has a minimum threshold of venture funding:

```
@cherrypy.expose
def job_small_town_rent_multiple(self,multiple=0.0,maxpop=999999999):
    jobs = con.query("""select ?job ?title ?salary ?location ?cid ?cname
                        where {?job jobs:vacancywith ?cid .
                            ?cid company:name ?cname .
                            ?job dc:title ?title .
                            ?job jobs:salaryrange ?range .
                            ?range jobs:minimumsalary ?salary .
                            ?job jb:location ?location .
                            ?cid jb:funding_round ?round .
                            ?round jb:amount ?amount .
                            filter(?amount > %d)
                            ?location jb:onebedrent ?rent .
                            filter (?salary > ?rent*%f)
                            ?location census:population ?pop .
                            filter (?pop<%d)}""" % \
                            (float(multiple), int(maxpop)))

    t = lookup.get_template('jobquery.html')
    return t.render(jobs=jobs, qp=quote_plus)
```

What else can you come up with? You could try writing a query that finds jobs at companies that are in different cities from their main headquarters. Or one that finds jobs that pay far too much as a percentage of their funding. You can also experiment with sorting and other SPARQL features. Hopefully this gives you a taste of what's possible when you connect a lot of different datasets in a semantic data store.

Visualizing the Job Data

We introduced Exhibit in Chapter 8, and it would be a shame not to make at least one visualization of our data. Let's do one that shows a scatterplot of salary versus funding and gives you lots of filtering options. Since Exhibit mostly operates via Ajax, the template for this page doesn't really have any dynamic elements. You can download the template from *http://semprog.com/psw/chapter10/templates/exhibit1.html*:

```
<html>
    <head>
        <title>Exhibit example</title>
        <link href="exhibit_job_data" type="application/json" rel="exhibit/data" />

        <script src="http://static.simile.mit.edu/exhibit/api-2.0/exhibit-api.js"
                type="text/javascript"></script>
        <script
         src="http://static.simile.mit.edu/exhibit/extensions-2.0/chart/
            chart-extension.js" type="text/javascript"></script>
    </head>
    <body>
        <table><tr><td>
          <div ex:role="viewPanel">
            <div ex:role="view"
                 ex:viewClass="Exhibit.ScatterPlotView"
                 ex:label="Funding vs. Salary"
                 ex:x=".funding"
                 ex:xLabel="Funding"
                 ex:y=".salary"
                 ex:yLabel="Salary">
            </div>
          </div>
        </td><td width="20%">
          <div ex:role="facet"
               ex:expression=".company"
               ex:facetLabel="Company"
               ex:height="10em"></div>
          <div ex:role="facet"
               ex:expression=".label"
               ex:facetLabel="Title"
               ex:height="10em"></div>
        </td></tr>
    </body>
</html>
```

The template specifies a page called `exhibit_job_data` as its data source, so we just need to create that method in our CherryPy application and pass data in a format that it recognizes:

```
@cherrypy.expose
def exhibit_job_data(self):
    jobs = con.query("""select ?job ?title ?salary ?amount ?cid ?location ?cname
                        where {?job jobs:vacancywith ?cid .
                               ?job dc:title ?title .
                               ?cid company:name ?cname .
                               ?job jobs:salaryrange ?range .
                               ?range jobs:minimumsalary ?salary .
                               ?job jb:location ?location .
                               ?cid jb:funding_round ?round .
                               ?round jb:amount ?amount .}""")
```

```
items = []
for row in jobs:
    items.append({'id':row['job']['value'],
                  'label':row['title']['value'],
                  'uri':'view?id='+quote_plus(row['job']['value']),
                  'funding':float(row['amount']['value']),
                  'company':row['cname']['value'],
                  'location':row['location']['value'],
                  'salary':float(row['salary']['value'])})

return dumps({'items':items})
```

Notice that we've used the Exhibit/JSON format rather than RDF. Either one would have worked, but since JSON is Exhibit's native format, it doesn't require the translation step, making it easier to test your code locally.

Finally, we need to render the template, which is easy since it doesn't have any dynamic content:

```
@cherrypy.expose
def exhibit(self):
    t = lookup.get_template('exhibit1.html')
    return t.render()
```

And we'll get a page that looks like what you see in Figure 10-9.

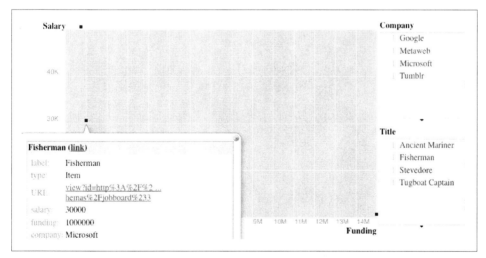

Figure 10-9. Screenshot of company data shown in Exhibit

There's a whole lot more you could do with visualization using Exhibit. For example, you can create maps with circles that correspond to a value, so you could visualize salaries across the country. We discussed timelines in Chapter 8, which would let you plot funding events for various companies over time. See what you can come up with!

Further Expansion

Since you can keep adding data to the application, it's easy to imagine a lot of other views and searches you might create. With more and more data coming online, there's really no limit to the directions you could take this. For example:

- In addition to demographic data about cities, you could add crime or school district data as a way to help determine the desirability of a city.
- You could find average salary ranges or "typical responsibilities" for the job titles.
- There's a lot more data about the financial state of companies that you could add to the companies themselves.
- People are always commenting on companies online, and there are sites dedicated to what it's like to work there. You could add links to commentary on various sites.

We encourage you to explore the different ways that companies and places tie in to the larger ecosystem of data. We also encourage you to try building an application just like this around your personal interests, whether you're into video games, rare birds, deepwater dives, or celebrities. If you create something cool, we'd love to hear from you!

Epilogue

The Giant Global Graph

Hopefully, over the course of this book we have demonstrated the power of semantic programming and convinced you of how easy it is to incorporate semantic technologies into your own architectures. Our ability to concisely introduce semantic technologies has been facilitated by a lot of existing research on building the semantic web over the past decade. And that existing work rests on more than a quarter century of research on language understanding and knowledge representation.

Today's semantic technologies are the result of threshing many lines of academic research that deal with a wide variety of thorny issues, and harvesting, milling, and packaging the most useful and practical solutions. This sifted and codified experience gives you a standard, flexible approach to data integration and information management whether you are writing a Ruby on Rails "mash-up" or an Enterprise Java industrial solution.

Research into semantic technology continues to this day. Many academic labs are actively pursuing various open questions of knowledge representation. However, this ongoing research doesn't mean the existing methods aren't ready for adoption. To the contrary, it means that your investment in semantic technology today is backed with continuing support. What you learn today will prepare you for the continuous and incremental deliveries emerging from this vibrant area of work. That said, it is important to sort out what is useful today from what may be useful in the future.

When you are considering a new semantic technology, look to see if it is embodied in multiple tools, whether a community of practice is emerging around it, and if the W3C has released any Recommendations about it. All of these are good indications that incorporating a prospective technology into your own architecture is a good idea.

Vision, Hype, and Reality

The semantic web was first widely introduced into technical vernacular with the publication of Tim Berners-Lee, James Hendler, and Ora Lassila's article "The Semantic Web" (*http://www.sciam.com/article.cfm?id=the-semantic-web*) in the May 2001 issue of *Scientific American*. The article outlined existing work on almost all of the foundational concepts covered in previous chapters, including triples, RDF, ontologies, and the role of URIs. The article also attempted to motivate the role of the semantic web by introducing an agent scheduling usecase, which used semantically annotated information to find health care providers that fulfilled a number of constraints.

In the years following the article, the promise of the semantic web has continued to grow both inside and outside the technical community. Mainstream publications such as *Newsweek*, *Forbes*, and the *New York Times* have covered various promises and prospects for semantically enabled data on the Web. But while many of the technologies necessary for building the semantic web were known in 2001, we have yet to see anything like the semantic web or the agents envisioned in the *Scientific American* article emerge.

The semantic web, however, is not a single thing, and as such it won't come into existence on any given day. Rather, the semantic web is a collection of standards, a set of tools, and, most importantly, a community that shares data to power applications. Unfortunately, throughout this period the W3C has continued to present the semantic web as a monolithic "stack" of technologies (see Figure 11-1). Sometimes referred to as the "layer cake," the stack illustrates how different technologies build upon one another, extending semantic capabilities. With all the technologies in place, user applications can be supported on top of the stack. Over the years, the layers of the cake have been refined to reflect the scope of evolving standards and to account for capabilities required by new usecases.

This view of semantic technologies reinforces the notion that the semantic web is a "thing" that does or does not exist. If this view of the semantic web were correct, then we are a long way off from being able to provide semantic web applications, as most of the top-level technologies identified by the stack are only fledgling research activities and have no standards in development.

But, as we have hopefully shown in this book, useful semantic applications using the nascent web of data can be built today. The view of the semantic web as a monolithic architecture misses the general point that semantic technologies are now sufficiently mature that the core methodologies have been codified in well-designed components that can be included in any application architecture. Many businesses are currently applying the practical semantic technologies that we have outlined in the book.

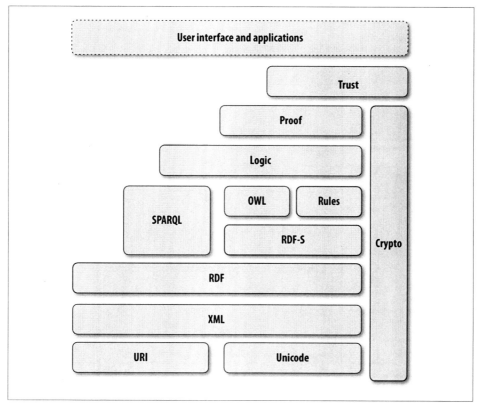

Figure 11-1. W3C semantic web technology stack

Figure 11-2 breaks the semantic stack into existing technologies that are already widespread (URIs, Unicode, XML); practical and proven semantic technologies that are being actively used by companies adopting semantic technologies (RDF, RDFS, SPARQL, and parts of OWL); and emerging research that exists mostly within the laboratory or in very proprietary business settings. There is an enormous amount of value that can be captured by bringing the middle layer of existing semantic technology into your software and your company today. At this point, the standards are settled, the software APIs are solid, and the community is growing. In the future, as the research questions around RDF-based rule systems and logic reasoners become settled and the standards emerge, you may be able to build the applications envisioned in Tim Berners-Lee's original vision of the Web. But there is no reason not to reap the benefits from the existing work now.

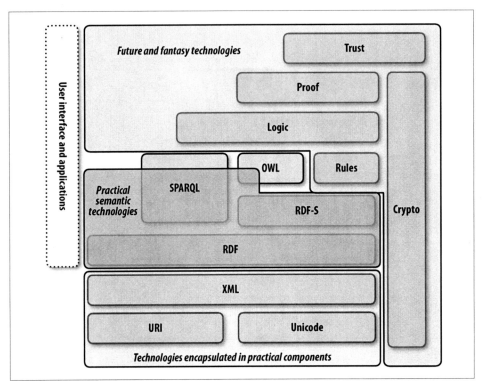

Figure 11-2. A practical view of the semantic stack

Participating in the Global Graph Community

In late 2007, Tim Berners-Lee, father of the World Wide Web, suggested that the phrase "semantic web" doesn't fully convey the granularity of the connections in his vision (*http://dig.csail.mit.edu/breadcrumbs/node/215*). "The Web" as we know it today is a network of interconnected documents, giving anyone the opportunity to publish their work and connect it to existing documents. The vision of "the semantic web" is a network of interconnected facts about real entities, published as graphs of data. Just as the existing Web is a graph of documents, the Giant Global Graph is envisioned to be a graph of graphs.

Anyone can publish their data into this massive joined graph, and anyone can connect their data to anyone else's data. The grand vision is one of all human knowledge offered up in a decentralized network of interconnected datasets. This sounds like an idealistic goal, but the whole of the World Wide Web today proves that it is possible to build interconnected systems on this scale, given the right technology and community.

As we have seen, the use of strong identifiers and shared semantics makes the vision of a Giant Global Graph possible. But the vision also requires the development of a global data community. With the knowledge you now possess, you are well positioned to play an important role in the Global Graph Community.

Releasing Data into the Commons

Whether you run a website or a business, it is easy to think of all of your data as an asset—as something that produces business value and therefore should be guarded like any intellectual property. However, if you try to carefully differentiate your company's value from your competitor's, it often becomes apparent that there is only a small amount of data that is truly valuable and that provides your competitive advantage. The vast majority of data is often simply the glue that connects your valuable data to the rest of the world and makes it possible to use.

This distinction can be described as follows. The things that produce business value and allow a business to differentiate are what Geoffrey Moore, a technology business analyst, calls "core." Everything else that goes on in a business that is not a part of value differentiation is deemed "context." Moore points out that to optimize business value, organizations should focus on those things that are core and treat everything that is context as a commodity service, outsourcing to lower-cost providers whenever possible.

From this perspective, the data from which your company derives value—the information about your users, the data produced through your R&D, the data that your competitors don't have access to—becomes core. All the other data that is used to make sense of the core data—geographic metadata used to organize your sales force, lists of industry standards that you need to comply with, databases of vendors that you keep on hand for convenience, product databases that everyone in your market must maintain—all of this is really context. Money and time spent maintaining and growing core data helps grow the company. On the other hand, resources spent on maintaining context data are overhead and operating costs—potentially a liability, and best if minimized.

By making data available to the Giant Global Graph, organizations have an opportunity to share the burden of maintaining context data by creating a commons of data. Shared data is a public good that lowers everyone's overall cost of doing business. And unlike other types of public goods, there can be no tragedy of the commons with information because information it can never be used up. If one group takes more than they contribute, the overall community is not hurt. There is only upside as people contribute into the commons.

License Considerations

Public data is only useful if you understand how you are allowed use it. Copyright is a complicated area of law, and it is often unclear if there are limitations on the use of public data. Traditionally, the metadata about permitted uses of information has been provided in legal documents intended for humans—and even then, comprehensible by only a few of us. For a truly global graph to work, machines must not only be able to discover data and join it to existing information, but they must also be able to understand how the data can be used.

Creative Commons (*http://creativecommons.org*) is a nonprofit organization that provides standardized ways to grant copyright permission to creative works. To convey the rights granted by information providers to data consumers in a machine-readable way, Creative Commons has developed the Rights Expression Language (ccREL), an RDF vocabulary for describing licenses. The vocabulary can be broken into two sets of predicates: Work Properties, which attach license statements to creative works, and License Properties, which describe the legal aspects of a license. See Figure 11-3.

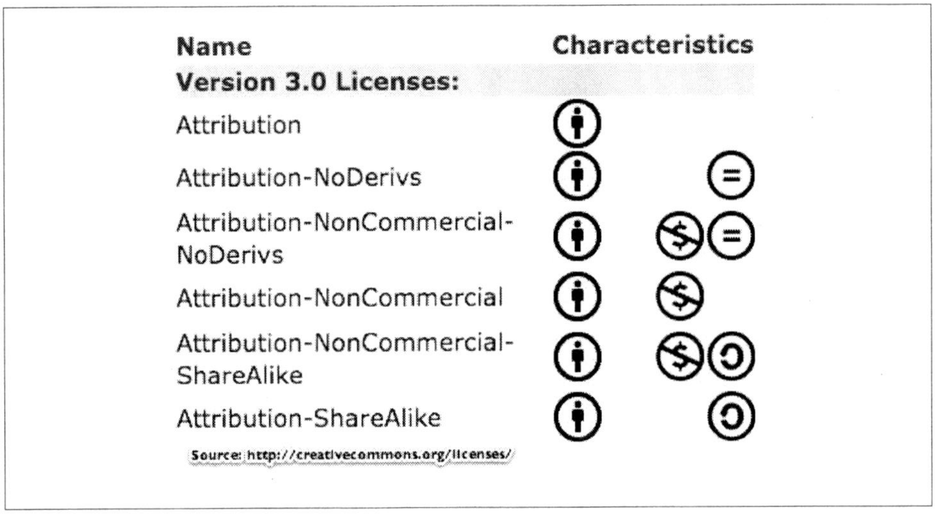

Name	Characteristics
Version 3.0 Licenses:	
Attribution	
Attribution-NoDerivs	
Attribution-NonCommercial-NoDerivs	
Attribution-NonCommercial	
Attribution-NonCommercial-ShareAlike	
Attribution-ShareAlike	

Source: http://creativecommons.org/licenses/

Figure 11-3. Creative Commons licenses

Creative Commons provides an array of pre-constructed licenses, which make use of the License Properties to describe how data can be used. To use Creative Commons licenses, you utilize the Work Properties to specify which resources are associated with which license. For instance, to grant rights so that anyone can use the MovieRDF dataset as long as they give attribution to Freebase, the source of the data, we make use of the XHTML license predicate to connect the resource for the dataset to the Creative Commons Attribution license. We then use Work Properties to indicate the name and URL that should be used for making proper attribution to the data creator. In N3, these RDF statements would be:

```
@prefix cc: <http://creativecommons.org/ns#>
@prefix xhtml: <http://www.w3.org/1999/xhtml/vocab#>

<http://semprog.com/psw/chapter4/movierdf.xml>
        xhtml:license <http://creativecommons.org/licenses/by/3.0/>
        cc:attributionName "Freebase - The World's database" ;
        cc:attributionURL <http://www.freebase.com/> ;
```

You can find additional predicates for specifying other information about the licensed work and the full set of License Properties for constructing new licenses at *http://wiki .creativecommons.org/CcREL*.

When you choose a license, think carefully about the restrictions that you place on your data. While selecting a license that disallows commercial usage or derivative work may seem like a conservative and prudent business decision, this often results in a lack of adoption of the data that you are trying to share. It can even lead to competing efforts to open up the same data with fewer restrictions, and can result in a community back-lash against your good intentions.

If you're trying to engage with a community of users and businesses, the Creative Commons Attribution license (CC-BY) is a liberal license that allows users and busi-nesses to use your data as long as they credit you when and where it is used. It allows your users and partners to freely build on your data, while ensuring that nobody claims that your data is their sole property.

The Data Cycle

When thinking about semantic data ecosystems, it is tempting to partition the world into data providers and data consumers. At first glance, it seems like these types of distinctions could help identify the technologies used by participants in different parts of the data ecosystem. Consumers of web pages use web browsers, while producers of web content use content management systems, application servers, and web servers. Hopefully we have demonstrated a symmetry in the consumption and production of semantic data. A consuming application will typically have a local triplestore in which to join semantic data from multiple sources, and will use graph queries to extract the information used by the application. A producing application will use a local triplestore on which to run graph queries, which are serialized out for specific requests. Both types of participants in the system use the same machinery, and both can consume and pro-duce data equally easily.

For this reason, networks of semantic data are more than point-to-point exchanges of information. Semantic applications are about integrating multiple sources of data through easy, standardized patterns. A good consuming application can reveal things that no singular source of its data "knew." The application can then republish that information in a machine-readable form for consumption by other applications.

This approach, where consumers are also producers, forms a data cycle. Information is transformed, augmented, and refined by a loose coupling of services that can be joined end-to-end by other applications. Each service transmits data in a standardized way, packaging metadata with data for use by other downstream services. Through the use of well-known vocabularies and license attribution, services further down the chain may have no contact with services earlier in the chain, but will know precisely how the data can be used. This vision of machines helping one another to help humans has long been a vision in science fiction, but with growth in the Linked Open Data community and applications that implement the trivial "serving" functions we have demonstrated, we are on the cusp of making this a reality.

Bracing for Continuous Change

RDF and semantic technologies are generative meta-models. They are frameworks that allow you to create languages for data interchange and integration. Their advantage is their extensibility. Rather than hardcoding data formats from multiple sources in your application and building custom integration code, semantic technologies provide a standardized toolkit and templates for specifying these operations. Making use of these techniques, even when the data sources are not semantically enabled, will allow you to adapt to new data and extend functionality without duplicating effort or working around fragile code.

Semantic technologies are not static, either—they continue to expand in capability. But because they are based on meta-models, extensions don't disrupt existing uses of the technology. When standardized rule expressions emerge, they will be expressed using the existing modeling language, and the systems that consume them will operate on the same model as well. In this way, an investment in semantic technologies is a hedge on future change—not only will your investment continue to yield, but you will have taken an option on using future technologies as they emerge.

The revolution of semantic technology is far from complete—we still have a long way to go. More semantic programming patterns will emerge, the tools will improve, and more people will learn the economies of semantic data. But hopefully we have convinced you that useful and powerful applications can be built *today*. There is a community of data providers who are motivated to share and curate large bodies of knowledge in a wide selection of domains, and you are welcome to join in that effort. Semantic technologies have arrived. They have been well incubated, and are now mature enough for serious application development.

We are truly interested in hearing what you are doing with semantic technologies, and we encourage you to join the semantic data community. Our FOAF information can be found throughout this book, and we have established *http://www.semprog.com* as a community resource for demonstrating practical approaches to semantic technology. We hope you will join the community and contribute new patterns, useful vocabularies, lessons learned, and best practices.

Though we are at the very early stages of the revolution, the tools for change are upon us. And as a meta-programming model, you are empowered to extend the framework for your expressive needs. After all, you don't have to wait for someone to invent the `<blink>` tag—you can do it yourself.

Index

Symbols

curly braces { }, with MQL, 123
dollar sign ($), SPARQL query variables, 86
question mark (?), SPARQL query variables, 86

A

about attribute (RDFa), 77, 158
add() method, 24
addFile() method (Sesame), 187
adding triples to stores, 24
    movie data example, 28–29
addString() method (Sesame), 187
_addToIndex() method, 24
addURI() method (Sesame), 187
AllegroGraph triplestore, 205
annotating XHTML pages (see RDFa)
anonymous nodes (see blank nodes)
application server, setting up (example), 232–234
applyinference() method, 44
artificial intelligence, 50
ASK queries (SPARQL language), 91–92
attributes, RDFa, 77, 158
autocomplete widget (Freebase), 125
average_clustering analysis (graphs), 103

B

backward chaining reasoning, 116
Bacon, Kevin, 51–53
BBC (British Broadcasting Company), 111
beer ontology, 149–152
behavior-oriented programming, 195
Berners-Lee, Tim, 64

betweenness_centrality analysis (graphs), 102
binding variables, 38–40
    implementing with triplestores, 40–43
blank nodes, 66
    in RDF/XML notation, 74
    representing in N3 notation, 72
    when describing people, 108
BOSS (Build Your Own Search Service), 105
breadth-first crawling (FOAF files), 98
breadth-first searches, 50–53
British Broadcasting Company (BBC), 111
browsing Linked Data, 179
business data
    graphs for, 33–35
    merging with place data, 53–54

C

camel-casing, 133
celebrity data, 31–33
    (see also movie data)
    Six Degrees of Kevin Bacon game, 51–53
centrality of graph nodes, 102
chains of inference rules, 47, 244
CherryPy framework, 232
Class class (OWL), 136
classes (in data models), 129–131
    film data model example, 132–134
    naming, in ontologies, 133
classes, OWL, 136
    comparing, 139
    disjoint, 146
classification-based inference, 43
cliques (graphs), 103
clouds of data, 106
clustering, graph nodes, 103

We'd like to hear your suggestions for improving our indexes. Send email to *index@oreilly.com*.

graph databases, 225
graphs
    blank nodes in, 66, 72
        to describe people, 108
        representing in N3 notation, 72
        representing in RDF/XML notation, 74
    examples of, 29–35
    of friends, 69
    merging, 26–28, 53–55
    persistent, 83–84
    searching for connections, 50–53
    social networks, 101–104
    visualizing, 55–58
        job listings application (example), 255–257
        viewing query results, 57
        viewing sets of triples, 56
Graphviz package, 55
Graphviz, for beer ontology, 149–152
Guha, Ramanathan, 64
GUIDs, Freebase, 118

H

hash (fragment) identifiers, 109
hCalendar event microformat, 157
hCard microformat, 156
href attribute (RDFa), 78, 159
hResume microformat, 157
HTML templates, creating (example), 234–237, 244–247
HTTP redirect to information resources, 108
human input, processing, 125–126
hype of semantic web, 262–263

I

iCalendar interchange format, 157
identifiers for blank nodes, 66, 72
    to describe people, 108
    representing in N3 notation, 72
    representing in RDF/XML notation, 74
identity, distributed approach to, 107
immutability of Freebase content, 118
in-memory triplestores (see triplestores)
indexes, 23
inference from triples (see feed-forward inference)
InferenceRule class, 44
inferencerule.py file, 44

infrastructure of Web, standardization of, 5
initNS keyword, query() method (RDFLib), 92
intelligence (artificial), 50
Internet Video Archive (IVA), 162–167
    RDFLib to Linked Data, 173–180
introspecting objects from data, 215–225
inverse functional properties (OWL), 146
inverse properties (OWL), 146
isomorphic() method (Graph class), 81
IVA (Internet Video Archive), 162–167
    RDFLib to Linked Data, 173–180

J

JavaScript Object Notation (JSON), 122
Jena toolkit, 204
job listing application (example), 227–258
    step 1: loading datasets into Sesame, 228–231, 251–253
    step 2: setting up web server, 232–234
    step 3: creating HTML templates, 234–237, 244–247
    step 4: expanding with public data, 237–244
    step 5: republishing for other applications, 248–251
    step 6: querying across datasets, 253–255
    step 7: using Exhibit visualizations, 255–257
joining graphs, 53–55
    querying joined graphs, 54–55
JSON (JavaScript Object Notation), 122
judgmental inference, 43

L

legacy data, 162–172
    Internet Video Archive (IVA), 162–167
        RDFLib to Linked Data, 173–180
    relational data, 169–172
    table and spreadsheet data, 167–169
license considerations, 266
LIMIT modifier (SPARQL queries), 95
link tags, for Exhibit, 208
Linked Data, 105–116
    browsing, 179
    consuming, 110–116
    outputting as, from RDFLib, 172–180
Linking Open Data (LOD), 106
literal values, in RDF, 68

load() method, 24
loading data into Sesame
    additional datasets (example), 251–253
    initial dataset (example), 228–231
    from public sources (example), 237–244
LOD (Linking Open Data), 106

M

_maketriples() method, 44
make_foaf_graph() method, 100
Mako templating language, 233
MCF (Meta Content Framework), 64
meaning, 19–35
    building schemas, in real world, 152–153
    building triplestores, 23–26
    deriving from triples (see feed-forward
        inference)
    flexibility of semantic data, 59
    graph representation examples, 29–35
    graph visualization, 55–58
        job listings application (example), 255–
            257
        viewing query results, 57
        viewing sets of triples, 56
    merging graphs, 26–28
    ontology and, 128, 148
    searching for connections, 50–53
MediaWiki, 169
MemoryStore class (Sesame), 185
merging graphs, 26–28, 53–55
    querying joined graphs, 54–55
Meta Content Framework (MCF), 64
metadata, as data, 16
metaweb.py library, 123–125
microformats, 156–158
Microsoft Excel spreadsheets, 6
    legacy data in, 167–172
modeling data, methods for, 5–14
    (see also data modeling)
    evolving and refactoring schemas, 9–11
    flexible schema extensions, 12–14
    relational data, 7–9
    tabular data, 6
    very complicated schemas, 11
movie data, 21–22
    (see also celebrity data)
    data model for, 132–134
    Internet Video Archive (IVA), 162–167
        RDFLib to Linked Data, 173–180

OWL schema definition for, 137
    querying using SPARQL, 84–96
    RDFObject example, 215–217
    representation of (example)
        adding and querying data, 28–29
        Six Degrees of Kevin Bacon game, 51–53
Movies API (Internet Video Archive), 162
MQL interface for Freebase, 121
MQL query language, 121
Mulgara triplestore, 204
MusicBrainz music metadata project, 111

N

N-Triple serialization, 70
N3 serialization, 72–73
Namespace class (RDFLib), 82
namespaces for URIs, 66
naming classes and properties, in ontologies,
    133
NetworkX package, 101
new triples, inferring, 44–45
nodes (graphs), analyzing, 101–104
normalization, flexibility versus, 13
.nt files, 71

O

object-oriented (OO) programming, 129
ObjectProperty class (OWL), 136
objects, introspecting from data, 215–225
objects, viewing with HTML templates
        (example), 234–237, 244–247
objects of triples, 19
    RDFa notation, 76–80
    serialization formats, 68–76
_objects() method (RDFObjectGraph), 221
ODE (OpenLink Data Explorer) extension,
    179
OFFSET modifier (SPARQL queries), 95
On-To-Knowledge project, 183
online services, 43
    geocoders, 46
ontologies, 127–153
    building real-world schemas, 152–153
    as data model, 128
    deciding what to exclude, 148
    existing, examples of, 148–152
    introspecting objects from data, 215–225
    language for (see OWL)

About the Authors

Toby Segaran is the author of *Programming Collective Intelligence (http://oreilly.com/catalog/9780596529321/)*, a very popular O'Reilly title. He was the founder of Incellico, a biotech software company that was acquired by Genstruct. He currently holds the title of Data Magnate at Metaweb Technologies and is a frequent speaker at technology conferences.

Colin Evans combines machine learning and semantic analysis into a deadly one-two punch against information entropy and noisy data. The results of his efforts appear as millions of facts in Freebase. Prior to joining Metaweb, Colin helped users organize their worlds through his work on the IRIS semantic desktop project at SRI.

Jamie Taylor started one of the first ISPs in San Francisco while developing an Internet laboratory for studying economic equilibria. This resulted in a better Internet connection at home and a Ph.D. from Harvard. He finally got a real job as CTO of DETER-MINE Software (now a part of Selectica), helping to create order in the unstructured world of Enterprise contract management. He is now helping to organize the world's structured information at Metaweb, where he oversees data operations.

Colophon

The animal on the cover of *Programming the Semantic Web* is a red panda (*Ailurus fulgens*, or "shining cat"), named for its bright, cinnamon-colored fur. Its resemblance to other animals has given it many nicknames, including firefox, cat bear, and lesser panda. Although predominantly red, it has white markings on its face and a black belly and limbs. Its long tail is ringed with red and yellow. Slightly larger than a domestic cat, the red panda is 30 to 50 inches long and weighs about 12 pounds.

Most wild pandas live on the slopes of the Himalayas and in the southwest forests of China. Preferring high altitudes, they move slowly on the ground but are agile climbers and foragers in the canopy. They subsist almost entirely on bamboo. During the hottest part of the day, the pandas sleep in shady treetops with their tails wrapped around their heads, becoming active only in the evening.

The red panda is endangered, due to habitat destruction, and is now a protected species in Nepal and China. No official numbers exist, but the total population is estimated at 2,500 individuals world-wide—mostly in Asian zoos—and rapidly declining. Contributing to their endangerment is the fact that red pandas are solitary creatures with a low birth rate and a high death rate in the wild, although their lifespan in captivity is 10–12 years. To counteract the population decline, zoos around the world have begun breeding programs, and several have successfully released captive-born pandas into the wild.

The cover image is from *The Riverside Natural History*. The cover font is Adobe ITC Garamond. The text font is Linotype Birka; the heading font is Adobe Myriad Condensed; and the code font is LucasFont's TheSansMonoCondensed.